GROUNDWATER CONTAMINATION FROM HAZARDOUS WASTES

GROUNDWATER CONTAMINATION FROM HAZARDOUS WASTES

PRINCETON UNIVERSITY WATER RESOURCES PROGRAM

Eric F. Wood, Director

Raymond A. Ferrara
William G. Gray
George F. Pinder

PRENTICE-HALL, INC., Englewood Cliffs, New Jersey 07632

Library of Congress Cataloging in Publication Data

Main entry under title:

Groundwater contamination from hazardous wastes.

Bibliography: p.
Includes index.
1. Water, Underground—Pollution—United States. 2. Hazardous wastes—Environmental aspects—United States. 3. Hazardous waste sites—Environmental aspects—United States. I. Wood, Eric F. II. Princeton University. Princeton Water Resources Program.
TD223.G76 1984   628.4'4564   83-10948
ISBN 0-13-366286-1

Editorial/production supervision and interior
  design: Tom Aloisi
Cover design: Ray Lundgren
Manufacturing buyer: Tony Caruso

© 1984 by Prentice-Hall, Inc., Englewood Cliffs, New Jersey 07632

All rights reserved. No part of this book may be reproduced, in any form or by any means, without permission in writing from the publisher.

Printed in the United States of America

10 9 8 7 6 5 4 3 2 1

ISBN: 0-13-366286-1

PRENTICE-HALL INTERNATIONAL, INC., *London*
PRENTICE-HALL OF AUSTRALIA PTY. LIMITED, *Sydney*
EDITORA PRENTICE-HALL DO BRASIL, LTDA., *Rio de Janerio*
PRENTICE-HALL CANADA INC., *Toronto*
PRENTICE-HALL OF INDIA PRIVATE LIMITED, *New Delhi*
PRENTICE-HALL OF JAPAN, INC., *Tokyo*
PRENTICE-HALL OF SOUTHEAST ASIA PTE. LTD., *Singapore*
WHITEHALL BOOKS LIMITED, *Wellington, New Zealand*

CONTENTS

Preface xi

1

AN INTRODUCTION TO THE PROBLEM 1

References 5

2

FATE OF HAZARDOUS CHEMICALS IN THE ENVIRONMENT 6

2.1. Introduction 6
2.2. Sources for Exposure to Hazardous Chemicals 8
2.3. Exposure, Risk, and Hazard Assessment 9
2.4. Estimating the Fate of Hazardous Chemicals 13
2.5. Transformation Processes for Mathematical Models 20

    2.5.1. Volatilization 20
    2.5.2. Photodegradation 21
    2.5.3. Biological Degradation 21

2.5.4. Chemical Degradation 21
2.5.5. Sorption 21
2.5.6. Ecological Interactions 23

References 24
Appendix 1: Values for the Exponential Function 26
Appendix 2: Definition of Concentration 27
Appendix 3: Reduction Rates 28

3

NUMERICAL SIMULATION OF GROUNDWATER CONTAMINATION 30

3.1. Overview 30
3.2. Role of Hydrology in Groundwater Model Development 31
3.3. Use of Physical Principles in Model Development 31
3.4. Terminology and Description of Flow 32
3.5. Terminology and Description of Transport 34
3.6. Conservation Principles 36
3.7. Conservation of Fluid Mass in a Volume 39
3.8. Darcy's Law and the Groundwater Flow Equation 42
3.9. Conservation of a Chemical Species 45
3.10. Example of Application of Groundwater Flow Equation 49
3.11. Summary 51

References 52

4

MONITORING OF HAZARDOUS WASTE SITES 53

4.1. Introduction 53
4.2. Conceptual Framework for Monitoring 54
    4.2.1. General Monitoring Objectives 54
4.3. Monitoring Network Design 59
    4.3.1. Basic Data for Monitoring Design 59
    4.3.2. Example 1: The South Brunswick Level 2 Network Design Problem 61

4.3.3. Design Considerations for Level 3 Monitoring Networks 63
 4.3.4. Example 2: Design of a Level 3 Monitoring Network for Estimating Model Hydraulic Conductivities and Boundary Conditions 67
 4.3.5. Example 3: Expansion of an Existing Monitoring Network 68

4.4. Surface Geophysical Monitoring Techniques 70

 4.4.1. Determining Stratigraphic Boundaries 70
 4.4.2. Determining Leachate Plumes 71

4.5. Operation of a Monitoring Network 71
4.6. Closing Comments 74

References 74

5

CASE STUDIES 75

5.1. A Case Study of Landfill Leachate Groundwater Contamination 75

 5.1.1. Site Information and Characteristics 76
 5.1.2. Mathematical Modeling 81
 5.1.3. Evaluation of Remedial Schemes 82

5.2. A Case Study of a Waste Disposal Pond Contaminating Groundwater 86

 5.2.1. The Contamination Problem 86
 5.2.2. Subsurface Characteristics 89
 5.2.3. Mathematical Modeling 89
 5.2.4. Sensitivity Analysis and Model Prediction 93

5.3. A Case Study of Accidental Groundwater Contamination 95

 5.3.1. The Contamination Problem 95
 5.3.2. Subsurface Characteristics 97
 5.3.3. The Origin and Extent of Contamination 99
 5.3.4. Remedial Measures 100

5.4. Discussion of Case Studies 103

References 105

6

DECISION FRAMEWORK FOR SITING HAZARDOUS WASTE FACILITIES 106

6.1. Introduction 106
6.2. The Decision Process: From Siting Criteria to Sites 108
6.3. The Overall Characteristics 108

    6.3.1. Appropriate 108
    6.3.2. Formal 109
    6.3.3. Qualitative and Quantitative 109
    6.3.4. Involve the Public 111
    6.3.5. Well Explained 111

6.4. Process of Site Selection 112

    6.4.1. Define Objectives 114
    6.4.2. Determine Attributes 117
    6.4.3. Modeling the Problem 118
    6.4.4. Review Results and Iteration 118

6.5. Making a Final Decision 120

References 120

7

APPLICATION OF DECISION ANALYSIS FOR SITING HAZARDOUS WASTE FACILITIES 121

7.1. Introduction 121
7.2. Sussex County Siting Study 122

    7.2.1. New Jersey Geology: Sussex County 122
    7.2.2. Identifying Candidate Sites 124
    7.2.3. Specifying the Objectives and Attributes 125
    7.2.4. Evaluating Site Impacts 126
    7.2.5. Evaluating Alternative Sites 129
    7.2.6. Comparing Alternative Sites 130
    7.2.7. Case-Study Summary 132

7.3. Converse/Tenech Siting Study for Sussex County 133
7.4. The Conjoint Approach to Siting 141

    7.4.1. Comparison to Decision Analysis 143

7.5. Closing Remarks 143

References 144

APPENDIX A

HYDROGEOLOGIC PARAMETERS FOR SITING LAND EMPLACEMENT FACILITIES   145

A.1.   Introduction   145
A.2.   Geology   146

    A.2.1.   Local Topography   146
    A.2.2.   Quality of Host Rock   148

A.3.   Soil   152

    A.3.1.   Transport Capacity   152
    A.3.2.   Sorption Capacity   155

A.4.   Hydrology   156

    A.4.1.   Proximity to Groundwater Supplies   156
    A.4.2.   Proximity to Surface Water Supplies   158

A.5.   Climate   158

References   158

INDEX   161

PREFACE

Chemical contamination throughout the environment is characteristic of our advanced technological society. Industrial production often generates by-products that are of little economic value but that, as residual waste material, may have severe environmental consequences. One of the primary pathways for migration of these contaminants is the water cycle. Groundwater contamination, particularly from hazardous wastes, has been recognized in recent years as a very serious national problem.

In order to fully understand the risks posed by landfills or by toxic waste spills, the public needs an increased awareness not only of the *consequences* but also of the *mechanisms* of groundwater contamination. It is our feeling that a suitable book on this topic does not exist at the introductory level. This book is a response to this need.

The book is appropriate for environmental geology/geography and environmental science courses at an introductory level. It is also appropriate for industrial engineers and managers who are responsible for management of hazardous waste and need to know more about groundwater contamination.

After an overview of the problem in Chapter 1, the next four chapters focus on the mechanics, evaluation, and analysis of groundwater contamination. Chapters 6 and 7 focus upon the siting of new hazardous waste landfills. Case studies show how the approaches are being applied in the field. For those readers with a limited background in geology, Appendix A describes some of the hydrological characteristics that are important in the evaluation of landfill sites.

This book carries as the author the Princeton University Water Resources Program. The original contributions were written by either faculty, staff, or students in the Program, as follows: faculty, Raymond A. Ferrara, William G. Gray, George F. Pinder, and Eric F. Wood; staff, Walter F. Saukin (also Associate Professor of Civil Engineering at Manhattan College, Riverdale, New York); and students, Jeffrey M. Bass, currently with Arthur D. Little, and Catherine McVay, current with ICOS Corporation of America, New York. The contributions of Mr. Bass and Ms. McVay were drawn from their Bachelor of Science in Engineering theses. Chapter 2 was authored by Raymond Ferrara, Chapter 3, William G. Gray; Chapter 4, Eric F. Wood; Chapter 5, edited by Raymond Ferrara from case studies put together by William Gray, George Pinder, and Walter Saukin with the assistance of Alice Gormley, a summer research assistant from Manhattan College; Chapter 6, Jeffrey Bass and Eric Wood; Chapter 7, Eric Wood, drawing upon the thesis of Catherine McVay for the case study results; and Appendix A by Catherine McVay. Raymond Ferrara and Eric Wood coordinated and edited the individual contributions into the (hopefully) unified book. Special thanks is due to Patricia Roman for her diligence in preparing the several drafts of the manuscript, and to Tom Agans for preparing the illustrations.

The financial support that enabled the preparation of the book came in part from the Andrew W. Mellon Foundation through its grant entitled "Hazardous Substances and Chemical Waste Disposal." The Water Resources Program was supported by this grant from August 1979 to July 1982; this support is gratefully acknowledged.

ERIC F. WOOD
Director, Water Resources Program
Department of Civil Engineering
Princeton University
Princeton, N.J. 08544

CHAPTER ONE | AN INTRODUCTION TO THE PROBLEM

In a typical metropolitan region, an abandoned canal ditch is used as a chemical dumping site for over thirty years, then closed. Several years afterward, homes are constructed adjacent to the site and an elementary school is opened. Two decades later, residents complain of noxious basement odors. The State Health Commissioner begins an investigation and soon declares a state of emergency, ordering the close of the public school and recommending evacuation of pregnant women and children under the age of two. Five days later, a federal emergency is declared and financial assistance is approved by the President of the United States. Plans are made for the evacuation of 238 families at government expense. It only now has become evident that a serious health hazard exists, the result of chemical waste disposal begun more than half a century earlier.

Until recently such an account might have been dismissed as highly improbable. Today we know this sequence of events to be only too factual, as they relate the history of the infamous Love Canal disaster. How many additional "chemical timebombs" may be present throughout this country and even throughout the world remains unknown. Love Canal has awakened this nation to the severe health implications of the improper disposal of hazardous and toxic wastes and the improper maintenance and management of disposal sites.

Classically wastes have been disposed of in ways that appeared to be the cheapest and the least repulsive to the public. Few people believed that serious consequences could ensue, even if the method being used was ultimately found to be inadequate. In most cases, the effects on the environment of the constituents of the wastes were not known. Furthermore, technology was not available that would make treating the wastes physically or financially possible for most industries. Very often the method of disposal simply involved land application, in one form or another.

Land disposal of hazardous wastes is a technique used by all sectors of the population. With the expansion of industry and the migration of a larger portion of the population to suburban and rural areas, the ability of land to absorb such wastes with no apparent consequence has diminished. The most catastrophic environmental impact has been the widespread contamination of the nation's groundwater supplies. Because a broad segment of the population depends upon this resource, the potential consequences are obviously grave.

The basic physical problem with land disposal of hazardous waste stems from the movement of water, originating as precipitation into and through the disposal site. The dissolution of waste material results in contaminants being transported from the waste site to larger regions of the soil zone and, too often, to an underlying aquifer. Problems of groundwater pollution frequently lead to the condemnation of wells and to the contamination of surface water bodies fed by the associated aquifer. In many instances, well contamination is not detected until years after land disposal of waste has begun, owing primarily to the slow movement of the conveying groundwater. For some chemical species, adsorption to soil particles may also act to retard the movement of the contaminant plume.

Not until the 1970s did the scope and severity of the problem become apparent. Action was initiated on local, state, and federal levels to solve problems caused by past disposal and to develop workable approaches to future disposal. Currently, a federally organized nationwide program is taking shape on regulation and disposal of various categories of wastes, often loosely referred to as hazardous wastes. With respect to past disposal, it is necessary to (1) locate current and former disposal sites, (2) describe the nature of those sites and their contents, (3) evaluate the environmental impacts of these sites, (4) determine legal responsibility, and (5) develop site-specific management plans and remedial schemes if determined to be necessary. A program regarding future hazardous waste disposal sites must concentrate on methods for (1) inventory of the quantity and quality of wastes being generated, (2) operation and management of disposal sites during their useful life, and (3) restoration of particular sites after disposal has ceased.

The location of future disposal sites is a major point of concern. Whether the general approach is to incinerate or reclaim waste, or a combination of the two, ultimately land disposal or storage will be involved. Thus, the siting of a landfill that will be secure in the future is of

AN INTRODUCTION TO THE PROBLEM

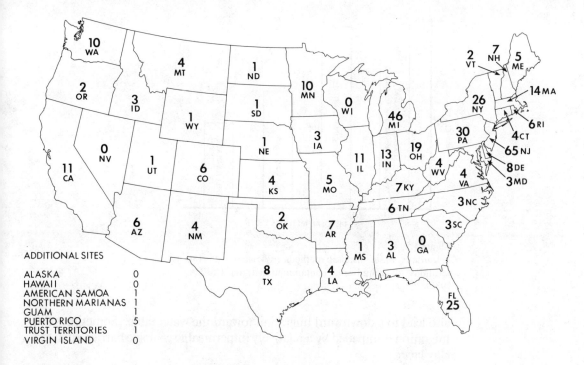

Figure 1.1.
Distribution of 418 Superfund priority sites, as of December 1982.

primary interest. The design and location requirements of such a landfill or disposal site are motivated by those of existing sites and their respective environmental impacts.

Figure 1.1 shows the distribution, as of December 1982, of the 418 priority hazardous waste sites (Superfund sites) as mandated under Section 105 of the Superfund law (Comprehensive Environmental Response, Compensation and Liability Act, PL96-510). The figure clearly demonstrates that groundwater pollution from hazardous waste disposal is not merely a local or regional problem but one that confronts the entire nation. Figure 1.2 presents, as a frequency diagram, the method of disposal associated with 180 contamination sites abstracted from USEPA (1980). It is evident that municipal landfills and surface disposal were the techniques that most frequently led to contamination problems. It was further observed in this study that disposal often was off the plant site.

Water pollution, caused by a hazardous waste facility, may evolve in a variety of ways. Leachate from landfills may drain out of the side of the fill and appear as surface runoff. Alternatively, it may seep down slowly through the unsaturated zone, if one exists, and enter an underlying aquifer. Breaks in liners of holding ponds or cracks in the bottom of storage tanks

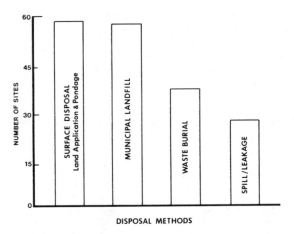

Figure 1.2.
Hazardous waste disposal methods associated with 180 locations of groundwater contamination (after USEPA, 1980).

also lead to a downward migration toward the water table. Sometimes this migration is impeded by a relatively impermeable geologic barrier, such as a clay layer.

Without the institution of remedial measures, buried waste usually acts as a continuing source of pollution. The waste constituents continue to be transported in the subsurface by infiltrating precipitation. Thus, it is generally recommended that sites involved with handling hazardous wastes be located above a natural barrier, as well as an applied liner. Moreover, the site should be instrumented to continuously monitor the condition of any associated aquifers. If leachate generation is anticipated, there should be a system for the collection and treatment of the leachate.

The Love Canal containment project provides an example of remedial action. Two principal barriers are placed in the way of migrating chemicals. Visible to the eye is the mounded clay cap that has been placed over the old canal. The clay has been compacted to maximize its resistance to seepage of rainwater into the wastes beneath. The cap is graded to divert stormwater into surface drains and, along with perimeter fencing, prevents direct human contact with potentially hazardous chemicals. The clay cap also helps to retain fumes from the volatile materials. Beneath the surface is an equally important part of the Love Canal containment: nearly 7,000 lineal feet of a three-foot-wide barrier drain filled with crushed stone and sand and extending 12 to 19 feet below ground level. A pipe at the bottom of this trench carries leachate from the dumpsite to a leachate treatment plant which removes the chemicals. This barrier drain isolates the canal from the surrounding environment. Since water has been the principal carrier of

Love Canal wastes to the surrounding environment, cutting off the flow of water beyond the barrier drain will also stem migration of the wastes. The drain is also expected to help cleanse the contaminated yards of former residences nearby, as water from abutting properties flows back into the drain and is treated.

In response to the growing public concern over the waste disposal problem, the federal government, on October 21, 1976, passed the Resource Conservation and Recovery Act (RCRA). This legislation sought to "promote the protection of health and the environment and to conserve valuable material and energy resources." These ends were to be accomplished through reclamation and disposal control. A portion of the law deals directly with the question of hazardous waste management, although some of the problems generated by abandoned or inactive waste disposal sites are not addressed. This section of the legislation suggests that the federal approach to the situation is to emphasize recovery rather than disposal. Public funds for remedial action related to prior mismanagement are to be provided through superfund legislation.

CHAPTER ONE REFERENCES

RESOURCE CONSERVATION AND RECOVERY ACT (Amendment to Solid Waste Disposal Act), Public Law 94-580, 90 Stat. 2795, 42 U.S.C. §6901 et seq. 1976.

UNITED STATES ENVIRONMENTAL PROTECTION AGENCY, *Damages and Threats Caused by Hazardous Material Sites,* Oil and Special Materials Control Division Report, 1980.

CHAPTER TWO | FATE OF HAZARDOUS CHEMICALS IN THE ENVIRONMENT

2.1. INTRODUCTION

The Resource Conservation and Recovery Act (RCRA) defines a hazardous waste as a solid waste that may cause or significantly contribute to serious illness or death, or that poses a substantial threat to human health or the environment when improperly managed. The United States Environmental Protection Agency (1980) has proposed that a hazardous waste be identified by testing it to determine if it possesses any one of four characteristics:

1. *Ignitability*—wastes that pose a fire hazard during routine management.
2. *Corrosivity*—wastes that require special containers because of their ability to corrode standard materials, or that require segregation from other wastes because of their ability to dissolve toxic contaminants.
3. *Reactivity*—wastes that tend to react spontaneously, to react vigorously with air or water, to be unstable to shock or heat, to generate toxic gases, or to explode during routine management.
4. *Toxicity*—wastes that may release toxicants in sufficient quantities to pose a substantial hazard to human health or the environment when improperly managed.

All of these characteristics produce *acute* effects likely to cause immediate damage. The fourth deals also with substances that tend to create *chronic* effects more likely to appear over a longer time period.

Table 2.1 Hazardous Waste Disposal in the United States[a]

Disposal Method	Percent of Total
Unacceptable methods:	
Unlined surface impoundments	48
Land disposal	30
Uncontrolled incineration	10
Other	2
Acceptable methods:	
Controlled incineration	6
Secure landfills	2
Recovered	2

[a] After USEPA (1980).

The danger to the environment is not from the production of the hazardous waste but rather from its ultimate disposal. Table 2.1 lists various disposal methods. Ninety percent of the total hazardous waste in the United States is disposed of by environmentally unsound methods. The reason is not that environmentally sound technologies for treatment and disposal of hazardous waste do not exist. A significant factor is cost, which can vary widely according to type and volume of waste handled. Naturally the costs for sound disposal practices are significantly greater (Table 2.2).

A primary mechanism for transport of improperly discarded hazardous waste through the environment is via the movement of water through ground and surface systems. The Love Canal episode cited previously is a clear example of hazardous waste migration through the groundwater system. Other episodes can be cited.

- Organic compounds leaching from a nearby landfill contaminated groundwater in the communities of Toone and Teague, Tennessee, rendering their drinking water supply unacceptable.
- Pesticide waste discarded in unlined surface ponds between 1943 and 1957 has contaminated groundwater in a 30-square-mile area near Denver.
- Fifteen hundred drums containing waste from metal-finishing operations buried near Byron, Illinois, contaminated ground and surface waters

Table 2.2 Cost of Hazardous Waste Disposal Practices[a]

Technology	Dollars per Metric Ton
Land spreading	2–25
Surface impoundment	14–180
Chemical fixation	5–500
Secure chemical landfill	50–400
Incineration (land-based)	75–2,000

[a] After USEPA (1980).

with cyanides, heavy metals, phenols, and other materials, destroying wildlife, stream life, and local vegetation.
- Seventeen thousand drums deposited on a 7-acre site in Kentucky, the "Valley of the Drums," contaminated surface waters with some 200 organic chemicals and 30 metals.
- Waste containing polychlorinated biphenyls (PCB's) discharged to the Hudson River has contaminated the riverbed sediments and rendered all fish unsuitable for human consumption.

2.2. SOURCES FOR EXPOSURE TO HAZARDOUS CHEMICALS

Section 2.1 briefly described several cases where hazardous chemicals contaminated the natural environment. The sources may be direct or indirect, intentional or unintentional, natural or artificial. We might also classify them into the following categories:

1. *Biological cycles*—including decay of animal and plant life, excretion of toxins, and so on.

2. *Domestic waste*—including discharges of raw or treated wastewater that contains, in addition to conventional pollutants, any of the hazardous compounds normally used in residential life and discharged via our sewage systems.

3. *Industrial waste*—again including raw or treated wastewater discharges.

4. *Nonpoint sources*—such as landfill leachate, septic tank leachate, and stormwater runoff containing hydrocarbons, solvents, fuels, oils, heavy metals, and so on.

The tragedy of the presence of polychlorinated biphenyls in our natural water systems provides a fine example of unintentional contamination. PCB's were formerly used by industry because of their resistance to heat. They are carcinogenic in laboratory rats and mice and are toxic to most life including humans. Their presence is so widespread that they have been found in animals 11,000 feet down in the North Atlantic. Indeed, the U.S. Environmental Protection Agency calculates that 91 percent of all Americans have detectable levels of PCB's in their adipose (fatty) tissue. Contamination is overwhelmingly apparent in the fish population of the Great Lakes, where health advisories have been issued because PCB levels in such fish are above the 5 parts per million (5 ppm) level recommended as acceptable by the Food and Drug Administration. A similar situation exists in the Hudson River, where for many years capacitor manufacturing plants were directly discharging PCB's to the river. In the Great Lakes atmospheric transport provides the major source of contaminants to the region, allowing precipitation to provide a direct input to the lakes.

Organic chemicals pervade many of our drinking water supplies. A well-known case involves an EPA-conducted monitoring program in the Mississippi River, which serves as a major water supply for New Orleans (EDF and Boyle, 1980). The study confirmed the presence of three chemical carcinogens—chloroform, benzene, and *bis*(2-chlorethyl)ether—and the EPA concluded that trace organics in the Mississippi were a potential threat to human health. Subsequently, an Environmental Defense Fund study concluded that there was a statistically significant correlation between drinking water from the Mississippi River and the occurrence of cancer in New Orleans residents.

Further studies by the EPA under their National Organic Reconnaissance Survey revealed widespread trace organic contamination of drinking water. Eighty cities were monitored. Chloroform was found in all eighty. Various other compounds also were identified. Miami exhibited the greatest contamination levels, with an observed chloroform concentration of 311 parts per billion.

Paradoxically, it appears that our very own water treatment and purification processes may be significantly contributing to the occurrence of carcinogenic substances in drinking water. Various epidemiological studies demonstrate increased cancer rates in populations consuming water supplies characterized by increased levels of chlorinated organic compounds (e.g., chloroform). In water treatment, the typical disinfection process used is chlorination. Low-level concentrations of organic compounds in the raw water can combine with chlorine to form chlorinated organic compounds. Thus, the treatment step that has been used so successfully to eliminate water-borne diseases due to bacterial infections may have some part in contributing to our low-level exposure to suspected carcinogenic compounds.

2.3. EXPOSURE, RISK, AND HAZARD ASSESSMENT

A basic question in our evaluation of hazardous wastes in the environment is that of the implications for human health. What risks are associated with consuming on a regular basis a certain amount of fish containing 5 ppm PCB? What are the risks at 1 ppm? Are the costs associated with banning the harvesting and sale of fish with these PCB levels worth the benefits? Questions such as these rely on the science of *risk assessment*. This technique quantifies the risks associated with different hazards and compares them with common risks. Humans are not now nor will they ever be completely free from risk. Clearly, there are limitations, financial or otherwise, to reducing those risks. As a result, the degree of severity of individual risks must be determined, so that the limited resources at our disposal can be expended to minimize the net risk. For example, simply stated, would government research funding be better spent improving the technology for disposal of hazardous wastes or improving medical and surgical techniques?

Risk assessment is characterized by four approaches or aspects (Great Lakes Basin Commission, August 1980). The most typical and best-known approach is *cost-benefit analysis,* which attempts to quantify for a given expenditure (i.e., cost) what gain (i.e., benefit) will be obtained. A second approach considers *revealed preferences,* which assumes that the choices people make in the marketplace demonstrate the risks they are willing to accept. A similar approach utilizes *expressed preferences*—preferences obtained by directly asking the population what level of risk is acceptable. A last approach suggests the use of natural standards to determine acceptable risks—that is, a particular activity should not increase risk above that which would occur naturally in the environment in the absence of the particular activity in question.

In order to complete an assessment of risk, we must first deal with the question of exposure. The *exposure concentration* of a particular substance is the concentration to which humans, fish, organisms at some level in the food chain, or even the environment as a whole are exposed at a particular time. Three basic elements are required to estimate exposure concentration (Haque et al., 1980):

1. Information on the source of the toxicant, including such items as production rate and release rate to the environment.

2. Characteristics of the toxicant that describe its ability to travel and react in the natural environment.

3. Data that can be used to estimate the population at risk, including occupational characteristics, medical surveillance data, and socioeconomic use habits.

The first two elements can be incorporated into mathematical models for estimating the transport and fate of the toxicant and therefore the concentrations that will appear in different sectors of the environment. The output of this exercise can be a time history of toxicant concentration at various locations in the environment. We can then compare this with previously defined "acceptable" concentration levels to note whether and how often such levels are exceeded. Such an approach can be useful in assessing, for example, the effects of pesticide application in agricultural areas.

This last step in exposure assessment actually coincides with the first step in hazard assessment—comparison of the toxicity of a chemical with the exposure concentration. The degree of hazard deals not only with toxicity but also with exposure. A highly toxic substance with no exposure is not hazardous, whereas a mildly toxic substance with high exposure could be very hazardous. Environmental toxicology generally assumes that (1) for pollutants occurring in the environment at low levels, toxicity is propor-

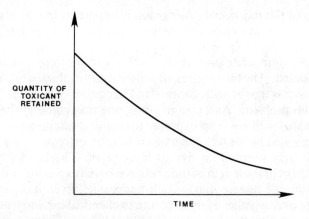

Figure 2.1.
Retention of toxicant over time.

tional to concentration, and (2) the longer an organism is in contact with a toxicant, the greater the probability of toxic effects, i.e., the greater the risk or hazard (Moghissi et al., 1980).

Exposure of any life form to a toxicant can lead to intake or absorption. The total amount of toxicant initially absorbed is gradually decreased via metabolic activity and excretion with other body by-products. In general, retention of a toxicant follows the pattern shown in Figure 2.1. The time integration of this function—i.e., the average concentration retained over a given time period multiplied by the length of time—is called the *retention dose*. The lifelong retention dose is called *dose commitment*. These doses may be expressed in units such as gram-days.

An equation for estimating the retention dose is (Moghissi et al., 1980).

$$R = C \times U \times D \qquad (2.1)$$

where
C = the concentration of the toxicant in the pathway medium (e.g., air, water, food),
U = the intake rate of the pathway medium for individuals in a given age group,
D = the retention-dose factor specific to the age and pathway medium, and
R = the annual retention dose of an individual of the age group through the pathway medium under investigation.

As can be seen, the retention dose is specific to the toxicant, the pathway of exposure, the concentration of the toxicant in the pathway, and the age group of the individual. Age group is important in that it represents food habits, excretion rates, and other physiological characteristics that determine retention of a toxicant.

Both immediate short-term as well as chronic long-term dangers must be considered. The former present a more easily evaluated problem in terms of the risk associated with contact and exposure. The latter present a more difficult problem. As a case in point, one may consider the potential risks associated with low-level exposure to chemical carcinogens. The problem is complicated by the fact that cancers have an extremely long latency period, which may range from five to forty years. The latency period may be thought of as the length of time between initial exposure to a carcinogen and the time the disease makes itself noticeable. Even more confusing is the debate over whether every exposure to chemical carcinogens presents some risk or whether there is a threshold level below which exposure presents no risk. The more conservative view that thresholds do not exist is embodied in the Delaney amendment to the Food and Drug Act, which prohibits the deliberate addition to foods of chemicals that have been shown to be carcinogenic in man or animals. A similar philosophy is embodied in the Federal Insecticide, Fungicide, and Rodenticide Act (FIFRA). The Ad Hoc Committee on the Evaluation of Low Levels of Environmental Carcinogens reported to the Surgeon General in 1970, "No level of exposure to a chemical carcinogen should be considered toxicologically insignificant for man" (EDF and Boyle, 1980).

The difficulty in identifying whether or not a threshold exists is related to our ability to design experiments capable of identifying that threshold level. Generally, animal carcinogenesis studies are conducted with high doses of carcinogens in relatively small numbers of animals. Inferences are then drawn from these results to predict what will happen when large numbers of humans are exposed to low doses of carcinogens. In other words, we must rely on extrapolation rather than interpolation of experimental data. Some scientists express the opinion that extrapolation of this sort is inexact and should not be used in a predictive sense.

In order to assess the effect of low-level exposure to humans we must rely on theories describing the relationship between exposure and response. The data is extrapolated in essentially two ways—one suggests that there is a threshold level below which no response is observed, the other that there is always some response regardless of how small the exposure level is—i.e., zero response exists only at zero exposure. The question as to which of these theories is correct may be simply academic, since zero exposure in today's technological society may be neither practically nor economically feasible. How then do we define a "safe" level? A common practice has been to assume that a safe level can be defined as one percent of the dose tested in animals.

In support of the threshold hypothesis, it has been observed that the latent period between exposure to the carcinogen and initiation of tumor growth generally increases as the dose is decreased. Researchers have found that the product of the dose and the latent time period raised to a power is a constant (Science, 1978):

$$K = d \times t^n \tag{2.2}$$

where

$K =$ a constant,
$d =$ the dose,
$t =$ the latent time period, and
$n =$ a constant power.

To more clearly demonstrate this relationship, this equation may be written as

$$t = \left(\frac{K}{d}\right)^{n'} \tag{2.3}$$

where

$$n' = \frac{1}{n} \tag{2.4}$$

Thus, the latent period is inversely proportional to the dose. This implies that some threshold level exists, since as d becomes very small (i.e., at small doses), t becomes large and may be several multiples of an individual's lifetime. Thus, the carcinogenic effect of the constituent has been effectively eliminated, owing to the magnitude of its "gestation" period.

Of course this approach deals with the carcinogenic effect of only a single independently acting constituent. The additive and even potentially synergistic effect of exposure to several chemicals at low levels is not well understood.

2.4. ESTIMATING THE FATE OF HAZARDOUS CHEMICALS

One problem in predicting the fate of toxic substances — or of any pollutant, for that matter — is that the science dealing with environmental pollution is so young, having surfaced formally only within the last two decades. The state of our knowledge, though growing daily, is limited. Unfortunately, decisions regarding regulation of environmental pollution need to be made today. When making decisions based on limited information, we must exercise caution, lest we introduce a danger that at some future date proves to be greater than the one we are eliminating.

A common tool used to make predictions is the *mathematical model,* generally a series of equations that attempt to describe the characteristics and response of some physical system. Mathematical models are used in many sciences. In economics, they are used to predict trends in market activity, the occurrence of recession or productive periods, and the like. In meteorology, they are used for prediction of short-term weather conditions as well as long-term climatological changes. In the environmental sciences, mathematical models can be used to predict the transport and fate of pollutants. Such models need not be complex sets of equations, incomprehensible to the average man, although in some cases these may be necessary. The advantage of mathematical models is that they enable the scientist and engineer to integrate in a consolidated framework all the various reactions that proceed simultaneously, affecting the concentrations of various pollutants.

The appropriate level of complexity for a model is probably the most controversial matter in any research effort. Of course, the true test of a model is its ability to make predictions that coincide with a previously measured set of observations. If this is done, the modeler may feel satisfied that he has chosen an appropriate structure for a particular case. However, one can never be certain that his choice is unequivocally better than any other. Each modeler will arrive at a particular model structure based on several objective and subjective tradeoffs. On one hand, a high level of complexity requires a sizeable number of rate coefficients and mathematical descriptions of transformation processes, which must be identified on the basis of a limited amount of knowledge. On the other hand, a simplified model, although requiring very few parameters, may give a poor conceptual view of the system and add little insight into the pertinent processes.

In formulating the model, we may use either of two basic approaches, "top-down" or "bottom-up." We begin the former by preparing a detailed conceptual model, which helps to organize otherwise unrelated facts and identify variable interactions that may not have been previously apparent. We can then make a series of refinements, trimming the model down to a size where all processes and input requirements can be reasonably described and the model output can be verified against an existing data base.

The bottom-up approach begins with the simplest mechanistic model (e.g., first-order decay of a total element) and then expands and adds detail until a point is reached where additional complexity does not result in significant output information (the point of diminishing returns). This method may be more appropriate in situations where data are severely limited or in cases where initial investigations are being made to identify trends of dominant characteristics of a system response.

Most mathematical models for the fate of pollutants have as their basis the conservation of mass. The equations define a *mass balance* or *material balance* for the pollutant in a given location. The equation may be written in words as

$$\begin{pmatrix} \text{the rate of} \\ \text{change of mass} \\ \text{at location } x \end{pmatrix} = \begin{pmatrix} \text{the rate of transport} \\ \text{of mass into and out} \\ \text{of location } x \end{pmatrix}$$

$$\pm \begin{pmatrix} \text{the rate of trans-} \\ \text{formation of mass} \\ \text{at location } x \end{pmatrix} \quad (2.5)$$

The physical space described by location x may be, for example, a section of a river, a lake, a coastal embayment, or a groundwater aquifer. It is commonly referred to as a *control volume* whose physical boundaries are defined. The transport term generally includes addition of mass due to inflows and the depletion of mass due to outflows from the control volume. The transformation term generally includes biological, chemical, and physical reactions that take place within the control volume. Examples may be decay of a radioactive substance, bacterial degradation of organic material, and so on.

Since the transport terms are primarily governed by fluid flows, they are described by the "hydraulic" properties of the system. This topic is dealt with later. Mass can also appear as a direct input to any control volume—for example, a direct waste discharge in concentrated form, which is then diluted and transported via the fluid flow in the system.

The transformation terms will be addressed in detail here. In general, we are dealing with changes in material over time. At equilibrium or steady-state conditions, the rate of change of mass is zero; that is, the transport and transformation terms are each of such a magnitude that they balance each other, so that an observer outside of the control volume sees no change. Such conditions do not always occur in the natural environment, where the system is said to be in a time-variable or dynamic state. The assumption of steady-state conditions is helpful in analyzing a problem in that it simplifies computations.

The transformation terms important in assessing the fate of toxic and hazardous substances are many. They include biological degradation, chemical degradation, adsorption/desorption, photodegradation, biological uptake, and volatilization. Before discussing these terms, let us look at their mathematical expressions.

Consider a case where a chemical has been accidentally discharged into a small pond for which inflow and outflow rates are negligible. The chemical is also known to decay at a constant rate, K_0. The rate of change of mass is then equal to this rate, and we may write

$$\begin{pmatrix} \text{rate of change} \\ \text{of mass } M \text{ in} \\ \text{the pond} \end{pmatrix} = \frac{\Delta M}{\Delta t} = -K_0 \qquad (2.6)$$

The symbol Δ describes a change and t denotes time. K_0 therefore describes a mass change per unit time—for example, pounds per day. The minus sign before the K_0 denotes a decay of mass. The solution is

$$\Delta M = -K_0 \cdot \Delta t \qquad (2.7)$$

If K_0 equals 2 pounds per day, we know that after one day, 2 pounds of M will have disappeared. In three days, 6 pounds will have disappeared. If the accidental spill originally included 10 pounds of M, then the pond will be free of the pollutant after five days.

The solution equation may be written in another way. If we define M_0 as the mass that was present at time zero—i.e., the mass accidentally discharged—then we may say the following:

$$\Delta M = M - M_0 \qquad (2.8)$$

or the change in mass over any time period equals the mass remaining at any time, M, minus the mass originally present, M_0. Since

$$\Delta M = -K_0 \cdot \Delta t = M - M_0 \qquad (2.9)$$

then

$$M = M_0 - K_0 \cdot \Delta t \qquad (2.10)$$

For $K_0 = 2$ lb/day, $\Delta t = 5$ days, and $M_0 = 10$ lb, we see that M equals zero.

In situations for which we are describing production or growth rather than decay, the equation would be written as

$$\frac{\Delta M}{\Delta t} = +K_0 \qquad (2.11)$$

with a solution

$$\Delta M = +K_0 \cdot \Delta t \qquad (2.12)$$

or

$$M = M_0 + K_0 \cdot \Delta t \qquad (2.13)$$

The equation above is a zero-order reaction which for now we can simply describe as one in which the rate of change of mass per unit time is a constant. A more common expression is the first-order reaction where the rate of change is proportional to the amount present at any time. For decay

$$\frac{\Delta M}{\Delta t} = -K_1 \cdot M \qquad (2.14)$$

for production

$$\frac{\Delta M}{\Delta t} = +K_1 \cdot M \qquad (2.15)$$

The solution is

$$M = M_0 \cdot \exp(-K_1 \cdot \Delta t) \quad \text{for decay} \qquad (2.16)$$

$$M = M_0 \cdot \exp(+K_1 \cdot \Delta t) \quad \text{for production} \qquad (2.17)$$

where exp ( ) is called the exponential function, some values for which are listed in Appendix 1 and plotted in Figure 2.2. Note that exp (α) equals 1 for $\alpha = 0$, approaches infinity as α becomes very large in the positive direction, and approaches zero as α becomes large in the negative direction. In these equations, K_1 is also a reaction rate coefficient, although it is somewhat different than K_0 discussed above.

For constant-volume systems the concentration, C, expressed as mass

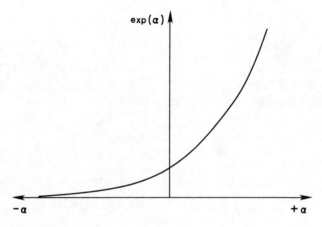

Figure 2.2.
Graphical representation of the exponential function.

per volume or mass per mass (see Appendix 2), may be substituted for the mass, M, in equations (2.14) through (2.17) (see Appendix 3).

A term frequently used to characterize the rate of decay of a pollutant in the environment is the half-life—the time after which the mass of a given substance will be one-half its initial value. For a first-order decaying substance this can be shown to be

$$t_{1/2} = \frac{.693}{K_1} \qquad (2.18)$$

The half-life is a function of the decay rate, K_1. Logically the greater the value of K_1, the shorter the half-life and the less long-lived the pollutant. For example, if K_1 has a value of 1 per day, then we can expect that one-half the pollutant will have decayed after .693 days or approximately $16\frac{1}{2}$ hours. But if K_1 is only .01 per day, then the half-life is 69.3 days. Substances that have very small decay rates, such as very stable organic compounds or radioactive isotopes, are therefore characterized by very large half-lives.

In many cases when a substance (e.g., A) decays, it produces an end-product (e.g., B) that is also of concern. In simple form

$$A \longrightarrow B \qquad (2.19)$$

If the equation expressing the rate of decay of A is written as

$$\frac{\Delta A}{\Delta t} = -K_1 \cdot A \qquad (2.20)$$

then the equation expressing the rate of production of B is written as

$$\frac{\Delta B}{\Delta t} = +y \cdot K_1 \cdot A \qquad (2.21)$$

where y is called a yield coefficient and expresses the mass of B produced per mass of A decayed. If for each unit of A that has decayed, one unit of B is produced, then $y = 1$ and the equation is simply

$$\frac{\Delta B}{\Delta t} = +K_1 \cdot A \qquad (2.22)$$

The solution is

$$B = B_0 + y \cdot A_0 - y \cdot A_0 \cdot \exp(-K_1 \cdot \Delta t) \qquad (2.23)$$

or

$$B = B_0 + y \cdot A_0 \cdot [1 - \exp(-K_1 \cdot \Delta t)] \qquad (2.24)$$

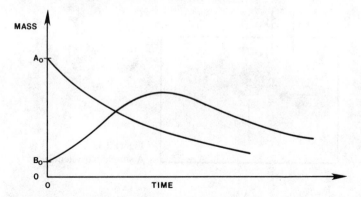

Figure 2.3.
Solution for a coupled system, A ⟶ B.

If B is also characterized by a first-order decay rate, e.g., K_1', then

$$\frac{\Delta B}{\Delta t} = +y \cdot K_1 \cdot A - K_1' \cdot B \qquad (2.25)$$

and the solution is

$$B = \frac{yK_1A_0}{K_1' - K_1}[\exp(-K_1 \Delta t) - \exp(-K_1' \Delta t)] + B_0 \exp(-K_1' \Delta t) \qquad (2.26)$$

where for convenience the multiplication signs have been dropped. Figure 2.3 demonstrates the solutions for A and B. Note that the solution equations become complex rather rapidly as the system complexity increases from one first-order decaying substance to two. For more complex systems involving several substances undergoing several reactions, direct analytical solutions as above are not possible and more sophisticated numerical mathematical techniques are necessary. Fortunately, with the proliferation and perfection of computers, such problems are becoming easier to solve.

The exercise above demonstrates the construction of a mathematical model, albeit a very simple one. The basic concepts presented are used by modelers in all professions. Consider the "ecosystem" in Figure 2.4 (Neely, 1980) into which some pollutant M is introduced at a rate k_0. The rate of addition may or may not be a constant. In the ecosystem the pollutant is dispersed. Some is taken up by fish (k_2) and some evaporates across the water surface to the atmosphere (k_1). The fish through normal metabolic activity excretes the pollutant at a rate k_3. If first-order reactions are assumed, the model equations can be written as

$$\frac{\Delta M_w}{\Delta t} = k_0 - k_1 \cdot M_w - k_2 \cdot M_w + k_3 \cdot M_f \qquad (2.27)$$

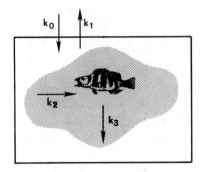

Figure 2.4.
A simple ecosystem.

$$\frac{\Delta M_f}{\Delta t} = k_2 \cdot M_w - k_3 \cdot M_f \tag{2.28}$$

where M_w and M_f express the mass of pollutant in the water and in the fish, respectively. By solving these equations, one would be able to estimate the amount of pollutant in the water as well as in the fish at any time for any input k_0. If we consider a case where the waterbody is a drinking-water reservoir, the fish are trout routinely harvested and consumed by men, and the pollutant introduced is PCB, the value of mathematical models for predicting the fate of hazardous substances becomes obvious. The models are used to estimate exposure concentrations, which in turn are used to evaluate the risk and hazard resulting from this level of exposure.

2.5. TRANSFORMATION PROCESSES FOR MATHEMATICAL MODELS

Although this chapter deals primarily with chemical contamination in groundwater systems, we will briefly discuss transformation processes for hazardous chemicals throughout the entire aquatic environment.

2.5.1. VOLATILIZATION

Many hazardous substances, particularly the organic compounds such as DDT and PCB, exist in a vapor as well as a liquid and a solid phase. When such substances are deposited in a waterway or on the land, a portion will evaporate or volatilize and travel through the air. Generalized mathematical relations have been developed to describe the transport of volatile compounds across the air-water interface, e.g., from the surface of a river or lake to the atmosphere. Volatilization can also occur from the solid (adsorbed) to vapor phases. This phenomenon is important in the soil environment.

2.5.2. PHOTODEGRADATION

Neely (1980) describes two main photochemical processes that account for chemical reactions in the environment: (1) direct photolysis, which involves the absorption of light followed by a transformation or degradation, and (2) a process involving a sensitizer molecule that causes an acceleration of the decomposition of chemicals. In the aquatic environment direct photolysis is important. Since light is absorbed and therefore attenuated as it passes through water, photolysis can occur only in the upper layers of a waterbody. The magnitude of the reaction will vary also with time of day, time of year, and latitude, all of which help determine the amount of incident sunlight at a particular location and time. The molecular structural characteristics of the pollutant in question are also important, as the energy required for destruction of a chemical bond will vary with the type of bond.

2.5.3. BIOLOGICAL DEGRADATION

Most degradation of organic compounds in the natural environment is carried out by microscopic organisms, primarily the bacteria. These microbes occur and are active in both the water and soil environments. Their ability to degrade specific compounds varies. Many synthetic organic chemicals are very resistant to microbial activity and are termed *refractory*. For these, the rate of introduction of the compound into the environment often exceeds the rate of degradation, and we therefore observe increasing concentrations throughout time until the rate of introduction is diminished.

2.5.4. CHEMICAL DEGRADATION

Various chemical reactions that occur in the environment lead to the destruction or neutralization of specific compounds. Of primary concern for this discussion is hydrolysis, a process whereby one functional group of a compound is replaced by another. An example is the hydrolysis of methyl bromide with hydroxide ion (Neely, 1980):

$$OH^- + CH_3Br \longrightarrow CH_3OH + Br^- \qquad (2.29)$$

The rate of such reactions depends on the chemical characteristics of the water medium.

2.5.5. SORPTION

Sorption is probably the major factor controlling the movement of many hazardous substances through subsurface systems and into groundwater aquifers. By *sorption* here we mean the physical/chemical bonding of some pollutant to a solid surface such as a soil particle. *Adsorption* refers to the attachment of the pollutant to the soil particle; *desorption* refers to the separation of the pollutant from the soil particle, freeing it to move in either the liquid or the vapor phase. The system strives toward attaining an equilibrium between adsorbed and desorbed phases based on the relative amounts of the pollutant in each of the solid, liquid, or vapor phases. We utilize a *partition coefficient, K_p*, to define the equilibrium point between any two phases. For example, the partition coefficient describing the equilibrium between DDT dissolved in groundwater and DDT adsorbed to soil may be written as

$$K_p = \frac{\mu g \text{ DDT adsorbed/g soil}}{\mu g \text{ DDT dissolved/l groundwater}}$$

where μg and g refer to units of weight, microgram and gram, respectively, and l is a unit of volume, namely liter.

In words, K_p defines the amount of DDT that will be adsorbed to soil particles for a given concentration of DDT dissolved in the groundwater. The partition coefficient defines the equilibrium condition, which may or may not be achieved because actual conditions are constantly changing. It is derived from an empirical equation known as the *Freundlich isotherm*, which states

$$S = KC^n \tag{2.30}$$

where

S = concentration in the adsorbed phase (mass of contaminant per mass of adsorbent),
C = concentration in the dissolved phase (mass of contaminant per volume of water), and
K, n = constants.

In many cases it has been found that the exponent, n, is equal to one, in which case S is linearly related to C, or $K = S/C$, which is the expression for the partition coefficient.

The higher the value of K_p, the less mobile is the pollutant, because for large values of K_p most of the pollutant remains stationary and attached to soil particles. In many cases it has been found that the degree of sorption is not a function of the soil itself but of the organic content of the soil. The phenomenon of adsorption is a primary reason why the sediment zones of our surface water systems are so highly contaminated with many hazardous organic compounds and heavy metals, and it also explains the slow rate of migration of contaminants through subsurface systems.

2.5.6. ECOLOGICAL INTERACTIONS

A major ecological problem of our time has been the increased concentration of chemicals in the higher forms of life. The sequential increase in concentration of a chemical in going from one trophic level to the next is known as *biomagnification* (Figure 2.5). The chemical is passed along the food chain in the natural predation process. Losses of the toxicant due to respiration and excretion at each level of the food chain are small. However, biomass decreases significantly, and as a result, although the mass of the toxicant is less as one proceeds up the food chain, its concentration (its mass per unit biomass) is increased.

At every level in the food chain various pathways exist for intake and removal. If we take the aquatic food chain specifically, pathways for intake to any level include direct uptake from the water, adsorption to the body from the water, and ingestion via a contaminated food source. Pathways for

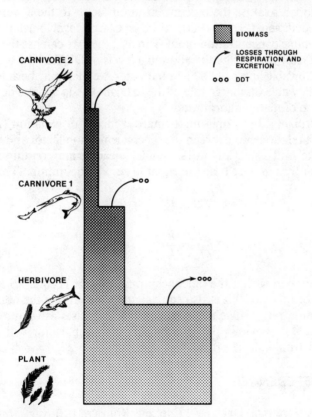

Figure 2.5.
Schematic illustration of DDT residues being passed along a simple food chain. Losses of residues along the chain are small in proportion to the amount transferred (after Neely, 1980; from Woodwell, 1967).

removal include respiration and excretion via normal metabolic activity and discharge of end-products, and normal death and decay.

The concept of partition coefficients as discussed previously under the section on sorption can be extended to describe the amount of substance that will appear in aquatic life. For example, we may define a K_p as

$$K_p = \frac{\mu g \text{ DDT in fish/g fish}}{\mu g \text{ DDT dissolved/l water}}$$

Used in this manner, K_p is the *bioconcentration factor* defining the amount of toxicant that will reside in fish for a given amount of toxicant in the waterbody. The term *partition coefficient* is reserved for an experimentally determined number that is the ratio of the equilibrium concentration of a chemical between two immiscible (nonpolar and polar) solvents. Octanol and water are generally chosen as the solvent pair. Equations have been developed relating the bioconcentration factor to the octanol-water partition coefficient (Neely, 1980). If we set a target level for what we consider an acceptable concentration of DDT in fish, then we can use the bioconcentration factor to indicate the allowable levels of DDT in the water. Knowing the allowable levels of DDT in the water, we then may be able to determine the allowable discharge rate of the toxicant to the aquatic environment via the use of mathematical models.

Utilizing the simple mathematical model of equation (2.28), we may note a relationship between the bioconcentration factor and the ratio of the uptake and excretion rates. Under steady-state conditions, changes in concentration over time are equal to zero by definition. Therefore

$$\frac{\Delta M_f}{\Delta t} = k_2 \cdot M_w - k_3 \cdot M_f = 0 \tag{2.31}$$

$$\frac{M_f}{M_w} = \frac{k_2}{k_3} \tag{2.32}$$

Remembering that the bioconcentration factor is equal to M_f/M_w, we see that for this model it is also equivalent to the ratio of the uptake rate to the excretion rate. Since the excretion rate decreases as we move up the food chain, it is obvious that the bioconcentration factor will be greater for upper-trophic-level organisms.

CHAPTER TWO REFERENCES

ENVIRONMENTAL DEFENSE FUND and ROBERT H. BOYLE, *Malignant Neglect,* Vintage Books Edition, May 1980.

GREAT LAKES BASIN COMMISSION, *Great Lakes Communicator,* Vol. 10, No. 11 (August 1980).

HAQUE, R., *Dynamics, Exposure and Hazard Assessment of Toxic Chemicals,* Ann Arbor Science Publishers, Inc., Ann Arbor, MI, 1980.

───── et al., "Role of Transport and Fate Studies in the Exposure, Assessment, and Screening of Toxic Chemicals," in R. HAQUE, ed., *Dynamics, Exposure and Hazard Assessment of Toxic Chemicals,* cited above.

MOGHISSI, A. A., et al., "Methodology for Environmental Human Exposure and Health Risk Assessment," in R. HAQUE, ed., *Dynamics, Exposure and Hazard Assessment of Toxic Chemicals,* cited above.

NEELY, W. B., *Chemicals in the Environment,* Marcel Dekker, Inc., New York, 1980.

SCIENCE RESEARCH NEWS, "Chemical Carcinogens: How Dangerous are Low Doses?" Vol. 202 (October 1978).

U.S. ENVIRONMENTAL PROTECTION AGENCY, "Everybody's Problem, Hazardous Waste," Office of Water and Waste Management, Washington, DC, 1980.

WOODWELL, G. M., *Scientific American,* Vol. 216, No. 24 (1967).

APPENDIX 1—VALUES FOR THE EXPONENTIAL FUNCTION

α	$\exp(\alpha)$	$\exp(-\alpha)$
0.00	1.000	1.000
0.25	1.284	0.7788
0.50	1.649	0.6065
0.75	2.117	0.4724
1.00	2.718	0.3679
1.25	3.490	0.2865
1.50	4.482	0.2231
1.75	5.755	0.1738
2.00	7.389	0.1353
2.25	9.488	0.1054
2.50	12.18	0.08209
2.75	15.64	0.06393
3.00	20.09	0.04979
3.25	25.79	0.03877
3.50	33.12	0.03020
3.75	42.52	0.02352
4.00	54.60	0.01832
4.25	70.11	0.01426
4.50	90.02	0.01111
4.75	115.6	0.008652
5.00	148.4	0.006738
5.50	244.7	0.004087
6.0	403.4	0.002479
6.50	665.1	0.001503
7.00	1097	0.000912
7.50	1808	0.000553
8.00	2981	0.000335
8.50	4915	0.000203
9.00	8103	0.000123
9.50	13360	0.000075
10.00	22030	0.000045

APPENDIX 2—DEFINITION OF CONCENTRATION

Assume a mixture of two substances, A and B. A definition of concentration is

$$C_A = \frac{M_A}{M_A + M_B}, \quad C_B = \frac{M_B}{M_A + M_B}$$

where C_A and C_B = concentration of A and B, respectively, and M_A and M_B = mass of A and B, respectively. Also:

$$\rho_T = \frac{M_A + M_B}{V_A + V_B}$$

where ρ_T = density of the mixture, and V_A and V_B = volume of A and B, respectively. In most water-quality problems we are dealing with dilute solutions where one constituent (water) is much greater in mass and volume than the other (pollutant). Assume that A (a pollutant) and B (water) are combined in a dilute solution.

$$M_B \gg M_A, \quad V_B \gg V_A$$

Then, $C_A \simeq M_A/M_B$, and $\rho_T \simeq M_B/V_B$ = density of water, which is 1.0 g/ml or 10^6 mg/l. Thus concentration is defined in units of mass per mass. However, for dilute solutions it is often expressed in units of mass per volume. For demonstration purposes assume a mixture consists of 1 mg A and 10^6 mg B or 1 part A per million parts B. Then we write

$$\frac{1 \text{ part A}}{10^6 \text{ parts B}} = 1 \text{ ppm} = \frac{1 \text{ mg A}}{10^6 \text{ mg B}} \equiv \frac{1 \text{ mg A}}{1 \text{ liter B}} = 1 \text{ mg/l}$$

Thus 1 ppm is equivalent to 1 mg/l with the designations A and B implied in the numerator and denominator, respectively. Similarly 1 part per billion = 1 ppb = 0.001 ppm = 0.001 mg/l = 1 µg/l. Therefore, in dilute solutions, concentration is often expressed in units of mass per volume according to the above.

APPENDIX 3—REACTION RATES

The rate of change of any substance for either a constant- or variable-volume system is written as

$$r = \frac{1}{V}\frac{dM}{dt} \tag{A3.1}$$

where

r = rate of change,
V = volume, and
M = mass.

Noting that the mass in the system at any time is equal to the concentration, C, times the volume, then

$$r = \frac{1}{V}\frac{d(CV)}{dt} = \frac{1}{V}\frac{V\,dC + C\,dV}{dt} = \frac{dC}{dt} + \frac{C}{V}\frac{dV}{dT} \tag{A3.2}$$

For a zero-order reaction

$$r = \pm\frac{K_0}{V} \tag{A3.3}$$

For a first-order reaction

$$r = \pm K_1 C \tag{A3.4}$$

Then for zero-order reactions

$$\pm\frac{K_0}{V} = \frac{dC}{dt} + \frac{C}{V}\frac{dV}{dt} \tag{A3.5}$$

$$\pm K_0 = V\frac{dC}{dt} + C\frac{dV}{dt} \tag{A3.6}$$

$$\pm K_0 = \frac{d(CV)}{dt} = \frac{dM}{dt} \tag{A3.7}$$

and for first-order reactions

$$\pm K_1 C = \frac{dC}{dt} + \frac{C}{V}\frac{dV}{dt} \tag{A3.8}$$

$$\pm K_1 CV = V\frac{dC}{dt} + C\frac{dV}{dt} \qquad (A3.9)$$

$$\pm K_1 M = \frac{d(CV)}{dt} = \frac{dM}{dt} \qquad (A3.10)$$

For constant-volume systems, $dV/dt = 0$ and equation (A3.5) becomes

$$\frac{dC}{dt} = \pm \frac{K_0}{V} \qquad (A3.11)$$

with solution

$$C = C_0 \pm \frac{K_0}{V}\Delta t \qquad (A3.12)$$

and equation (A3.8) becomes

$$\frac{dC}{dt} = \pm K_1 C \qquad (A3.13)$$

with solution

$$C = C_0 \exp(\pm K_1 \Delta t) \qquad (A3.14)$$

Note the similarity among the following equations:

(2.6) & (2.11) $\Leftrightarrow$ (A3.11)
(2.10) & (2.13) $\Leftrightarrow$ (A3.12)
(2.14) & (2.15) $\Leftrightarrow$ (A3.13)
(2.16) & (2.17) $\Leftrightarrow$ (A3.14)

CHAPTER THREE
NUMERICAL SIMULATION OF GROUNDWATER CONTAMINATION

3.1. OVERVIEW

The study of groundwater aquifers has traditionally and necessarily relied heavily on field data to provide insight into flow direction, water quality, or geologic formation properties. Because drilling costs are high and contamination problems complex, investigators have recently turned to numerical models as tools for insight into where data might best be collected or for comparing the effects of various well-pumping strategies on a particular aquifer. Output from such models is typically a computer-generated plot of some sort.

Society's faith in technology, and particularly in the computer as an objective and infallible instrument, has led to an attitude in some quarters that, for example, computer models of hydrologic systems are precise predictors of the fate of pollutants in the environment. In fact, simulators are based on principles of physics as well as on the hydrologic setting and available data. Deficiencies in any of these areas are directly reflected in inaccuracies in the model. Thus, to understand a numerical model, it is important to first understand the concepts and principles upon which the model is based. For the individual with no background in this field, this chapter provides an introduction to computer modeling of subsurface flow and chemical transport.

3.2. ROLE OF HYDROLOGY IN GROUNDWATER MODEL DEVELOPMENT

Hydrology is simply the study of water, and *groundwater hydrology* is concerned with water beneath the surface of the earth. Unfortunately, when studying a groundwater contamination problem, it is not possible to isolate the aquifer of interest from other components of the hydrosphere. For example, the atmosphere may provide rainwater that seeps into the ground and leaches contaminant into the aquifer. In considering this type of system, we need to know, at least to some degree, the amount of moisture available. Alternatively, a river may recharge an aquifer, in which case we may need to know the depth of water or stage in the river and the effect of a storm on the stage.

Although the interaction of atmospheric water and free-flowing water with groundwater is important, numerical models of groundwater flow are based primarily on groundwater hydrology. The interaction of groundwater with the surface water occurs around the boundaries of the groundwater reservoir. Thus typically we account for the interaction by specifying conditions along the boundary of the aquifer of interest.

With boundary conditions specified for a water-bearing formation, a hydrologic study becomes concerned with the physical and chemical interaction among the water, the contamination, and the geologic formation.

3.3. USE OF PHYSICAL PRINCIPLES IN MODEL DEVELOPMENT

A numerical model study of an aquifer requires that the qualitative information be supplemented by some basic laws of physics. These laws, which will be discussed in more detail subsequently, are conservation of mass, conservation of momentum, and conservation of a chemical species. These balance equations apply to portions of the aquifer as well as to the aquifer as a whole. By dividing the aquifer into subdomains and applying the conservation equations simultaneously to all subdomains, we can compute a hydrologic state for each subdomain. The more subdomains used, the higher the degree of resolution of the conditions in the aquifer—provided, of course, that the data can support the high resolution.

For example, if the concentration of a pollutant in an aquifer is known at only one point, it is difficult to infer what the pollutant distribution is in the aquifer. Further, it would be even more difficult to estimate how this unknown distribution would respond if the aquifer were pumped in a certain manner. Thus it would be little more than folly to analyze more than one subdomain of the aquifer (i.e., the aquifer as a single unit) for concentration. On the other hand, if concentration has been measured at a number of positions in the aquifer, such that the shape of a concentration plume is well defined, a high degree of discretization may be appropriate.

In the remainder of this chapter the basic hydrologic, physical, and numerical principles important to groundwater modeling will be introduced. Because the purpose is only to familiarize the reader with the methodology, the quantitative study will be limited to aquifers whose properties vary only in the two horizontal dimensions (i.e., two-dimensional aquifers). The study of vertical infiltration of water through the upper soil layer or of stratification within a formation is beyond our present scope.

3.4. TERMINOLOGY AND DESCRIPTION OF FLOW

Definitions in hydrology tend to be imprecise, with some relative qualitative descriptors being used that depend on the experience and needs of the investigator. Nevertheless, to be able to discuss the hydrologic properties of geologic formations, we need to have some fluency in the various terms. Figure 3.1 is a cross section of a coastal geologic formation that encompasses most of the important formation types.

First of all, geologic formations are differentiated according to their ability to bear and transmit water. Formations that have the ability to transmit and store water in economically usable quantities are referred to as *aquifers.* The *permeability* or *hydraulic conductivity* is a measure of the ease with which water is transmitted through the geologic formation. Aquifers with a high permeability typically consist of clean coarse sands or mixtures of sand and gravel. For such units the permeability varies roughly from .01 to 100.0 cm/sec. When considering flow that is virtually horizontal, we use the *transmissivity,* which is the product of the permeability and the thickness of the flow within the aquifer. A geologic unit that has virtually no transmissivity is called an *aquifuge.* A formation with very small transmissivity is an *aquiclude,* while one with a transmissivity between that of an aquiclude and an aquifer is called an *aquitard.* Although neither an aquiclude nor an aquitard transmits significant amounts of water in the horizontal direction, they are distinguishable in that an aquitard is important in transmitting water in the vertical direction. With reference to Figure 3.1, the formation between aquifers B and C behaves mostly as an aquiclude, but in the region labeled "leaky" it is an aquitard. Typical material for aquitards and aquicludes is clay, or silty clay with a permeability of .000001 to .01 cm/sec.

Because aquifers are the most important subsurface sources of water and are the primary formations capable of carrying a pollutant a significant distance, a number of adjectives are used to describe the type of aquifer under consideration. A *confined* aquifer is one bounded above and below by formations with permeability significantly less than its own. Further-

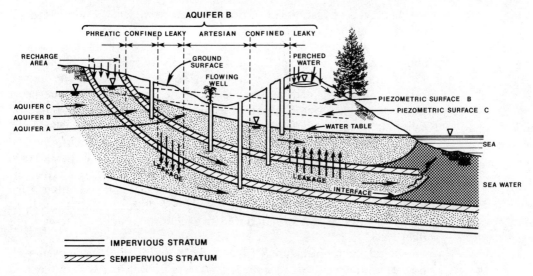

Figure 3.1.
Types of aquifers in a coastal geologic formation (after Bear, 1979).

more, if a well punctures the overlying layer and is screened in this aquifer, the water level in the well will be above the base of the confining layer. If either of the confining layers of the aquifer is pervious (e.g., an aquitard), the formation is referred to as a *leaky* confined aquifer. Aquifers B and C in Figure 3.1 are primarily confined aquifers, with leakage occurring through portions of their confining layers.

Aquifers that do not have an upper confining formation are referred to as *unconfined* or *phreatic* aquifers. The upper boundary of a phreatic aquifer is the water table. In some instances, pumping of the water from a confined aquifer can cause a decrease of the water-table elevation near the well such that this portion of the aquifer may behave as an unconfined aquifer. Recall from the description of confined aquifers that the water level in a well puncturing the top of the formation must be above the base of the confining layer. Wherever this criterion is not met, the aquifer is considered to be phreatic. In Figure 3.1 aquifer A is a phreatic aquifer.

When a well is drilled and screened in a phreatic aquifer in which there is no vertical flow, the water level in the well will be equal to the water-table elevation. The elevation in the well above some datum (usually mean sea level), is called the *hydraulic head* or simply *head*. If two observation wells or piezometers in a phreatic aquifer indicate different elevations of head, flow is occurring from the higher head toward the lower. In leaky aquifers where some vertical flow is occurring, the head will vary through the vertical, with flow direction again being from the higher head to the lower. In determining the direction of horizontal flow, it is important to consider the effect on head readings of vertical flow or screening of the piezometers at different elevations in the aquifer.

When wells tap a confined aquifer, the water levels in the wells will be above the upper surface of that aquifer. The surface defined by the water elevation in a number of wells in such an aquifer is called the *piezometric surface* (surface defined by water levels in piezometers). The water-level elevation above a datum in an observation well in a confined aquifer is called the *piezometric head*. The flow direction is from the higher head value toward the lower. In Figure 3.1, for example, the piezometric surface for aquifer B is higher than the water table of aquifer A, and thus leakage between them is from B into A, even though aquifer B is under aquifer A. On the other hand, the head in aquifer C is below that of B and thus leakage is from B into C. In the fresh-water portion of all three aquifers, the head decreases towards the right, and thus in each formation flow is toward the right.

A final feature of Figure 3.1 that deserves comment is the artesian well that taps aquifer B. Because the head of aquifer B is higher than the land surface at some points, water will flow freely from a piezometer whose top is below the piezometric surface but above the land surface. This is referred to as an *artesian condition*. Withdrawal of water from the other, nonartesian well requires pumping.

It has been mentioned that the direction of flow in a groundwater aquifer is from the higher head values to the lower. In fact, studies have shown that when velocities are small, as they usually are in groundwater systems, the velocity is directly proportional to the head gradient. The constant of proportionality is the permeability. Thus, if one is interested in modeling the velocity at which water will carry pollutant in an aquifer, it is essential to have accurate head measurements and permeabilities.

Permeability data typically are obtained from pump tests, whereby the lowering of the piezometric head throughout an aquifer due to pumping is monitored. A surveyor can provide accurate elevations of well casings above mean sea level, and an electric tape or another instrument is used to determine the water level in the well. A full discussion of field techniques is beyond our present scope; for further detail the interested reader should refer to a groundwater hydrology text.

3.5. TERMINOLOGY AND DESCRIPTION OF TRANSPORT

Vital to the study of any groundwater pollution problem is an understanding of the mechanisms for movement of the contaminant through the system. Thus before transport (i.e., contamination movement and decay) can be studied, it is imperative that the hydrology of the system be well understood and accurately parameterized. A prediction or simulation of contaminant movement depends upon quantities such as the permeability, porosity, and piezometric head. The error inherent in specifying these quantities is carried into the transport model and compounded. Thus, a precise description of a flow system is a prerequisite to any attempt to model pollutant movement.

This study will concentrate on a two-dimensional areal description of pollution transport modeling. Even when the flow is essentially two-dimensional, however, the contamination field may be three-dimensional. For example, if a contaminant is introduced into a phreatic aquifer by leaching of a landfill, one might expect pollution concentrations to be highest in the upper portion of the aquifer near the landfill. As the contaminant moves away from its source, it may mix through the aquifer thickness, and vertical stratification will disappear. In this case a two-dimensional model may prove adequate. On the other hand, concentrations may be higher near the bottom of the aquifer because of high pollutant density or due to infiltration of clean water from the land surface. In this instance it may be necessary to use a three-dimensional model or a multilayer two-dimensional model. Although the complexity required of the model must depend upon the particular system being studied, the mechanisms or physical processes that cause the pollutant movement are the same: convection, diffusion, dispersion, and reactions (including chemical processes as well as absorption and desorption).

Convection is the movement of pollutants at a particular velocity in a particular direction. Typically, a pollutant dissolved in water will be convected, on the average, at the same velocity and in the same direction as the water. This mechanism is generally considered to be the most important in pollutant movement. Thus the convection of a contaminant depends upon the groundwater flow velocity, which depends upon the piezometric head distribution, the permeability, and the porosity. When modeling, then, we need to determine the impact of errors in the latter quantities upon any predicted pollutant plume. We can do so by running a computer model with various values of these parameters to determine the sensitivity of the model to any errors.

Molecular diffusion is the mixing of a contaminant with water due to the random movement of the molecules. For example, if one puts some alcohol in a glass of water and lets the mixture sit at rest, the alcohol will mix, eventually, with the water, owing to the molecular motion or diffusion, although no net motion of fluid will occur. If one wished to use a more violent method of mixing, one could insert a rod into the glass and stir. Again there has been no net movement of the fluid (i.e., the glass of fluid is still in its initial location), yet significant mixing has occurred. This mixing occurs simultaneously with the diffusion process but completely overwhelms it. It is caused by mechanically creating velocity variations in the fluid, thus increasing the rate at which water and alcohol molecules come into contact. This mixing process is akin to *mechanical dispersion,* which is caused by local velocity variations within the aquifer induced by the complex geometry of the actual flow path. Diffusion, which depends on random molecular motion, exhibits no preferred direction of mixing, while dispersion, which depends on the mean flow velocity, is greater in the direction of flow than transverse to the flow. Typically, when a nonzero velocity field exists, diffusion is negligible compared to dispersion. Further, it should be obvious that in the absence of concentration gradients, both diffusive and dispersive mixing are zero.

The final mechanism that is important in describing pollutant transport is *reaction.* Some contaminants—for example, radionuclides—will decay (or be produced) over time by conversion to other species. Some species also interact with the soil and are absorbed onto the surface of soil particles. This mechanism is made use of in carbon filtration systems, which remove hydrocarbons from contaminated water. Although many of these reactions have been studied under laboratory conditions, transfer of lab information to field situations has proven to be most difficult. The reaction mechanism is extremely important for the accurate description of contaminant transport, but its complex parameterization has thus far led to rather crude incorporation of the process into numerical models.

3.6. CONSERVATION PRINCIPLES

Earlier sections in this chapter have described various physical features of the groundwater flow and transport process and introduced appropriate terminology for the discussion of this problem. To make use of this descriptive information in a numerical model, we must first quantify information using mathematical descriptions of various conservation principles.

The numerical modeler begins with differential equations that express conservation of mass, momentum, and chemical species. He then approximates these equations over the various subdomains being considered and solves these approximate equations. For the present case, differential equations are not considered to be necessarily within the reader's field of competence. Thus we will derive the necessary equations from a difference, rather than differential, point of view and show how they can be incorporated into a numerical model of an aquifer. The intent here is not to provide a totally rigorous development of the equations but rather to give the reader a feel for the concepts used in numerical modeling.

The basic rule of conservation invoked states that

$$\begin{pmatrix} \text{the rate of change} \\ \text{of a quantity in a} \\ \text{volume with time} \end{pmatrix} = \begin{pmatrix} \text{the amount of that} \\ \text{quantity which flows} \\ \text{into the volume per} \\ \text{unit time} \end{pmatrix}$$

$$- \begin{pmatrix} \text{the amount of that} \\ \text{quantity which flows} \\ \text{out of the volume} \\ \text{per unit time} \end{pmatrix} + \begin{pmatrix} \text{the amount of the quantity} \\ \text{added to the volume by} \\ \text{other means} \end{pmatrix} \quad (3.1)$$

In the derivations that follow, the volume to be considered is depicted in Figure 3.2. We will be considering two-dimensional flow, as mentioned previously, in the xy plane. The volume considered will be rectangular in the xy plane, with dimensions Δx and Δy in the x and y direction, respectively. The volume need not have a constant height, but it is assumed that some length, denoted b, is characteristic of the volume height. (Thus, it can be seen that the smaller the values of Δx and Δy, the more nearly this approximation will be satisfied.) We will also make use of the difference notation operator Δ in the following manner:

1. Δx, Δy refer to the length $x_{i+1} - x_i$ and $y_{j+1} - y_j$, respectively.

2. The difference between two different values of a quantity, say f, will be noted Δf.

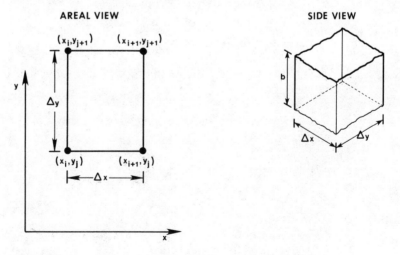

Figure 3.2.
Elemental volume used in derivation of the groundwater equations.

3. If we wish to indicate the change in value of f as we proceed across the element of volume in the x direction per unit change in x (i.e., the gradient of f in the x direction),

$$\frac{f_{i+1} - f_i}{x_{i+1} - x_i} = \frac{\Delta f}{\Delta x} \qquad (3.2)$$

where

f_{i+1} = the value of f at position x_{i+1},
f_i = the value of f at position x_i, and
$x_{i+1} - x_i$ = the distance between x_{i+1} and x_i.

For simplicity the i-subscript, which indicates the positions at which f and x are evaluated, has been dropped on the right side of this equation.

4. If we wish to indicate the change in value of f as we proceed across the element in the y direction per unit change in y (i.e., the gradient of f in the y direction)

$$\frac{f_{j+1} - f_j}{y_{j+1} - y_j} = \frac{\Delta f}{\Delta y} \qquad (3.3)$$

where

f_{j+1} = the value of f at position y_{j+1},
f_j = the value of f at position y_j, and
$y_{j+1} - y_j$ = the distance between y_{j+1} and y_j.

Again, for simplicity, we have dropped the subscripts in invoking the difference notion.

5. If we wish to indicate the change in the value of f per unit time change, we write

$$\frac{f_{t+\Delta t} - f_t}{\Delta t} = \frac{\Delta f}{\Delta t} \qquad (3.4)$$

where

f_t = the value of f at time t,
$f_{t+\Delta t}$ = the value of f at time $t + \Delta t$, and
Δt = the time interval between evaluations of f.

In the interest of simplicity, some of the more subtle aspects of the use of the difference notation have not been brought out. It is important to note, however, that if one is interested in evaluating or examining $\Delta f/\Delta x$, the notation is assumed to imply that one is keeping the y coordinate and the time coordinate fixed. Obviously, $\Delta f/\Delta x$, which is computed for one small volume, need not have the same value as $\Delta f/\Delta x$ for a volume located somewhere else in space.

3.7. CONSERVATION OF FLUID MASS IN A VOLUME

The most basic of the conservation equations is the conservation-of-mass or continuity equation, which, for the fluid in the volume of aquifer in Figure 3.3, states:

$$\begin{pmatrix}\text{rate of change of}\\ \text{mass of fluid in}\\ \text{a volume with time}\end{pmatrix} = \begin{pmatrix}\text{rate of flow of}\\ \text{fluid mass into}\\ \text{the volume}\end{pmatrix} - \begin{pmatrix}\text{rate of flow of}\\ \text{fluid mass out}\\ \text{of the volume}\end{pmatrix} \quad (3.5)$$

This equation, unlike the general equation (3.1), has no additional terms, because mass cannot be created within a volume. The next task is to account mathematically for all the terms in (3.5).

To develop (3.5) mathematically, we will have to make use of the following quantities:

ρ = density, the mass of fluid per unit volume of fluid. It has units of mass divided by length cubed or M/L^3.

u, v = the velocity of the fluid within the pore space in the x and y directions, respec-

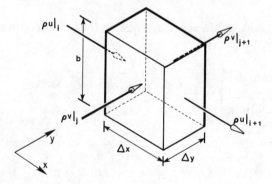

Figure 3.3.
Elemental volume with horizontal mass fluxes indicated.

tively. Velocity has units of length divided by time or L/T.

ϵ = the porosity of the material, the fraction of the volume considered that is not occupied by the solid grains. ϵ is also equal to the fractional areal surface of a boundary face of the volume that is not occupied by the solid material. Thus, ϵ is a measure of the fraction of the surface area across which fluid can flow, as well as a measure of the fraction of the volume available for fluid occupation.

The first term in (3.5) is a measure of the rate of change of fluid in the volume. Experimentally, one might measure this term by weighing the volume at two different times and determining the change in amount of fluid in the volume during the interval. Thus one must measure the mass per volume and multiply by the volume to obtain the mass, or

$$\begin{pmatrix} \text{fluid mass} \\ \text{in the volume} \\ \text{at time } t \end{pmatrix} = (\rho \epsilon b \, \Delta x \, \Delta y) \Big|_t \qquad (3.6)$$

where $\epsilon \, \Delta x \, \Delta y b$ is the volume occupied by the fluid. Then the fluid mass in the volume at time Δt later is

$$\begin{pmatrix} \text{fluid mass} \\ \text{in the volume} \\ \text{at time } t + \Delta t \end{pmatrix} = (\rho \epsilon b \, \Delta x \, \Delta y) \Big|_{t+\Delta t} \qquad (3.7)$$

The rate of change of mass is then equal to the difference in mass at the two times divided by Δt or

$$\begin{pmatrix} \text{rate of change of} \\ \text{fluid mass in the} \\ \text{volume with time} \end{pmatrix} = \frac{(\rho \epsilon b \, \Delta x \, \Delta y)|_{t+\Delta t} - (\rho \epsilon b \, \Delta x \, \Delta y)|_t}{\Delta t}$$

$$= \frac{\Delta(\rho \epsilon b \, \Delta x \, \Delta y)}{\Delta t} \qquad (3.8)$$

The next term that needs to be evaluated in (3.5) is the mass of fluid flowing into the volume. This is equal to the mass flow per unit area multiplied by the area. This term has three components: one that accounts for the x-direction inflow, one that accounts for the y-direction inflow, and one that accounts for net vertical recharge due to leakage from above and below.

$$\begin{pmatrix} \text{rate of flow of} \\ \text{fluid mass into} \\ \text{the volume} \end{pmatrix} = (\rho u \epsilon b \, \Delta y)\bigg|_i + (\rho v \epsilon b \, \Delta x)\bigg|_j + \rho R \, \Delta x \, \Delta y \quad (3.9)$$

where $\epsilon b \, \Delta y$ and $\epsilon b \, \Delta x$ are the areas the x and y components of flow cross, R is the recharge rate, and $\Delta x \, \Delta y$ is the area through which the recharge occurs. The outflow expression is similar to (3.9) except that terms are evaluated at $i+1$ and $j+1$ instead of i and j, respectively, and instead of net recharge, a term that accounts for pumping wells is included:

$$\begin{pmatrix} \text{rate of flow of} \\ \text{fluid mass out} \\ \text{of the volume} \end{pmatrix} = (\rho u \epsilon b \, \Delta y)\bigg|_{i+1} + (\rho v \epsilon b \, \Delta x)\bigg|_{j+1} + \rho Q \, \Delta x \, \Delta y \quad (3.10)$$

where $Q \, \Delta x \, \Delta y$ is the volume of water pumped per unit time from the volume.

Combination of the terms in (3.8), (3.9), and (3.10) according to equation (3.5) yields the mass-balance equation

$$\frac{\Delta(\rho \epsilon \, \Delta x \, \Delta y b)}{\Delta t} = (\rho u \epsilon b \, \Delta y)\bigg|_i - (\rho u \epsilon b \, \Delta y)\bigg|_{i+1} + (\rho v \epsilon b \, \Delta x)\bigg|_j$$
$$- (\rho v \epsilon b \, \Delta x)\bigg|_{j+1} + \rho R \, \Delta x \, \Delta y - \rho Q \, \Delta x \, \Delta y \quad (3.11)$$

Although ρ, ϵ, u, v, and b are assumed to vary spatially and temporally, such that, for example, these variables may have different values at i and $i+1$, Δx and Δy are constants. Thus, equation (3.11) may be divided by Δx and Δy to obtain

$$\frac{\Delta(\rho \epsilon b)}{\Delta t} = \frac{(\rho u \epsilon b)|_i - (\rho u \epsilon b)|_{i+1}}{\Delta x} + \frac{(\rho v \epsilon b)|_j - (\rho v \epsilon b)|_{j+1}}{\Delta y} + \rho R - \rho Q$$
$$(3.12)$$

Then, making use of the Δ notation defined in equations (3.2) and (3.3), we obtain

$$\frac{\Delta(\rho \epsilon b)}{\Delta t} = -\frac{\Delta(\rho u \epsilon b)}{\Delta x} - \frac{\Delta(\rho v \epsilon b)}{\Delta y} + \rho R - \rho Q \quad (3.13)$$

This is the basic form of the continuity equation used in modeling. Subsequently this equation will be modified, and the reader may have difficulty relating the final version back to the physically meaningful concept of conservation of mass. Nevertheless, if the basic physical premise and the progression of mathematical steps are correct, the final equation derived must also be correct.

One of the most commonly used modifications to (3.13) involves the *storage coefficient,* denoted by S. The storage coefficient is the volume of water that an aquifer will absorb (or release) from storage per unit planar area per unit change in head. The storage coefficient is therefore dimensionless. If we denote the head by ϕ and the change in head by $\Delta\phi$, the quantity $\rho S \Delta\phi$ is the mass of water that an aquifer absorbs (or stores) as a result of a head change of $\Delta\phi$ per unit area. As discussed previously, the only way that mass can be added to or subtracted from a volume is via convection or flow of the mass into or out of the volume. Thus the quantity $\rho S(\Delta\phi/\Delta t)$ must equal the left side of equation (3.13), and therefore

$$\rho S \frac{\Delta\phi}{\Delta t} = -\frac{\Delta(\rho u \epsilon b)}{\Delta x} - \frac{\Delta(\rho v \epsilon b)}{\Delta y} + \rho R - \rho Q. \tag{3.14}$$

In equation (3.14), S and ϵ are properties of the aquifer being considered while ρ is the water density. These must be specified. The parameter b is the thickness of a confined aquifer or the depth of flow in an unconfined aquifer (and thus directly related to the water-table elevation). The unknowns that must be solved for in (3.14) are the velocities, u and v, and the head ϕ.

3.8. DARCY'S LAW AND THE GROUNDWATER FLOW EQUATION

In the mid nineteenth century Henri Darcy, a French engineer, performed a series of classical experiments to investigate the behavior of flow in porous media. He found that in flow through a pipe packed with porous material, the flow per unit area is proportional to the head difference between the two ends of the pipe and inversely proportional to the pipe length. The constant of proportionality between the flow and the head gradient is the permeability, and the relation obtained by Darcy is

$$q = -K \frac{\Delta\phi}{\Delta l} \tag{3.15}$$

where

q = the volumetric flow per cross-sectional area,
K = the permeability,
$\Delta\phi$ = the head difference across the length of the pipe, and
Δl = the pipe length.

Equation (3.15), commonly referred to as *Darcy's law,* is not strictly a law but only a useful relation that applies to most (though certainly not all) naturally occurring flows in porous media. Among the effects that may cause non-Darcian behavior are high velocity or the presence of more than one liquid phase (such as oil flowing with water).

Although Darcy's law was developed from experimental evidence for one-dimensional flows in porous media, it has been extended to two- and three-dimensional flows. In systems that have the same permeability in all directions, called *isotropic media,* equation (3.15) is generalized so that the flow in any direction is proportional to the head gradient in that direction. For a two-dimensional system, such as the one that is our primary interest,

$$q_x = \epsilon u = -K \frac{\Delta \phi}{\Delta x} \quad (3.16a)$$

$$q_y = \epsilon v = -K \frac{\Delta \phi}{\Delta y} \quad (3.16b)$$

where q_x and q_y are the volume of flow per unit time per unit cross-sectional area in the x and y direction, respectively. The flow components q_x and q_y are called *Darcy velocities,* or *superficial velocities,* and are smaller than the actual average velocities in the pores by the factor ϵ.

For media that behave according to equation (3.16), the direction of flow is always the same as the head gradient direction. However, in some media, referred to as *anisotropic media,* this is not the case. For example, in a limestone formation, caverns may exist that follow one predominant direction. A head gradient not collinear with this direction may still give rise to flow through the caverns. In an anisotropic material, flow from one point to another moves along the path of least resistance and not necessarily along the straight line connecting the two points. Modeling of anisotropic flows requires that different permeabilities be specified in each direction and that terms be included that account for flow in one direction due to a gradient in an orthogonal direction. Thus, the anisotropic analogue to (3.16) would be

$$q_x = \epsilon u = -K_{xx} \frac{\Delta \phi}{\Delta x} - K_{xy} \frac{\Delta \phi}{\Delta y} \quad (3.17a)$$

$$q_y = \epsilon v = -K_{yx} \frac{\Delta \phi}{\Delta x} - K_{yy} \frac{\Delta \phi}{\Delta y} \quad (3.17b)$$

where

K_{xx} = the permeability in the x direction due to a head gradient in the x direction,

K_{xy} = the permeability in the x direction due to a head gradient in the y direction,

K_{yx} = the permeability in the y direction due to a head gradient in the x direction, and

K_{yy} = the permeability in the y direction due to a head gradient in the y direction.

Available data often are not extensive enough to permit modeling of anisotropy with any significance. For purposes of further discussions here, only the isotropic version of Darcy's law, as in equation (3.16), will be used. The reader should keep in mind, however, that directional differences in permeability, particularly between vertical and horizontal directions, can significantly influence a flow pattern and the rate of transport of contaminants.

The linear relation between velocities and head gradient as given in (3.16) makes it possible to eliminate the velocities from equation (3.14). Substitution of (3.16a) and (3.16b) into (3.14) for ϵu and ϵv results in

$$\rho S \frac{\Delta \phi}{\Delta t} = \frac{\Delta}{\Delta x}\left(\rho K b \frac{\Delta \phi}{\Delta x}\right) + \frac{\Delta}{\Delta y}\left(\rho K b \frac{\Delta \phi}{\Delta y}\right) + \rho R - \rho Q \quad (3.18)$$

Recall that Kb is equal to the transmissivity T, so that (3.18) becomes

$$S \frac{\Delta \phi}{\Delta t} = \frac{\Delta}{\Delta x}\left(\rho T \frac{\Delta \phi}{\Delta x}\right) + \frac{\Delta}{\Delta y}\left(\rho T \frac{\Delta \phi}{\Delta y}\right) + \rho R - \rho Q \quad (3.19)$$

If the problem under consideration is one for which the density does not vary spatially, then density may be divided out of equation (3.19) to yield

$$S \frac{\Delta \phi}{\Delta t} = \frac{\Delta}{\Delta x}\left(T \frac{\Delta \phi}{\Delta x}\right) + \frac{\Delta}{\Delta y}\left(T \frac{\Delta \phi}{\Delta y}\right) + R - Q \quad (3.20)$$

This balance equation is the form most commonly solved to obtain a head profile or the flow field in an aquifer. This equation states that per unit area, the rate of change of fluid volume is equal to the net inflow in the x direction plus the net inflow in the y direction plus the net recharge due to vertical leakage minus the net discharge due to pumping wells.

To model an aquifer, one divides it into a number of adjacent subvolumes and then applies equation (3.20) to each subvolume. This set of equations is supplemented by boundary conditions at the periphery of the aquifer that provide head values or flux conditions. Once this system is solved for the distribution of head, the velocity field may be computed using the difference form of Darcy's law in equation (3.16). The velocities are important, as they give an indication of the rate at which contaminants move, and they are used in the species balance equation to be developed in the next section.

3.9. CONSERVATION OF A CHEMICAL SPECIES

The equation to be developed here is similar to that for conservation of mass. The difference is that only one chemical species in the fluid mass is being followed. If one writes equations for each of the species being transported and then adds up all the equations, the result must be equation (3.14), which is the overall balance equation. In what follows, a conservation equation will be derived for only one species. To develop a complete model, one must derive equations for all species present and account for the chemical interactions among these species. Usually, the data base does not permit an analysis of this great detail, and the modeler must be content with describing one species or a representative contaminant.

For the following derivation, the contaminant that will be followed is referred to as "species A." The balance equation is

$$\begin{pmatrix} \text{rate of} \\ \text{change of mass} \\ \text{of species A} \\ \text{in a volume} \\ \text{with time} \end{pmatrix} = \begin{pmatrix} \text{rate of mass} \\ \text{flow of species} \\ \text{A into the} \\ \text{volume} \end{pmatrix} - \begin{pmatrix} \text{rate of mass} \\ \text{flow of} \\ \text{species A} \\ \text{out of the volume} \end{pmatrix}$$

$$+ \begin{pmatrix} \text{net rate at which} \\ \text{species A is produced} \\ \text{in the volume by} \\ \text{chemical reaction} \end{pmatrix} \quad (3.21)$$

In developing the mathematical statement of this principle, we will introduce the following notation:

ρ_a = the mass of species A per unit volume of fluid, having units of mass over length cubed or M/L^3.

R_a = the net rate at which species A is produced in the fluid, having units of mass per unit volume per unit time. R_a may be a complicated expression and depend on the concentration of other species in the fluid. If R_a is negative, chemical A is being consumed by the reaction.

J_{ax}, J_{ay} = the dispersive fluxes of species A in the x and y directions, respectively, with units of mass per unit length per unit time. The solution for the head profile in conjunction with Darcy's law gives the values of the average fluid velocity. However, different chemical species may move at different velocities, and J_a accounts for the difference between the velocity of species A and the average velocity of the fluid. The importance of this term can be realized if one considers a droplet of ink in a glass of water. If the ink and water particles all move at the same velocity, there can be no distributing of the ink through the water. However, simply stirring this system will yield a uniform mixture, because, in fact, the average velocity in the system is not the velocity for all particles.

As with the derivation of the mass-balance equation, the first term in (3.21) can be evaluated by measuring the mass of A contained in the volume of Figure 3.3 at two different times and using these measurements to determine the rate of change, or

$$\begin{pmatrix} \text{rate of change of} \\ \text{mass of species} \\ \text{A in a volume} \\ \text{with time} \end{pmatrix} = \frac{(\rho_a \epsilon b \, \Delta x \, \Delta y)|_{t+\Delta t} - (\rho_a \epsilon b \, \Delta x \, \Delta y)|_t}{\Delta t}$$

$$= \frac{\Delta(\rho_a \epsilon b \, \Delta x \, \Delta y)}{\Delta t} \quad (3.22)$$

The next term evaluated is the flow of A into the volume. This flow is due to convection, dispersion, and recharge. It is important that recharge into the volume may carry any concentration of A, but negative recharge, or leakage from the volume, must carry the concentration of A that is present in the volume. To stress this point, ρ_a' will be used to indicate this mass of A per fluid volume and is equal to ρ_a for leakage out but must be specified independently of ρ_a for recharge. Thus

$$\begin{pmatrix} \text{rate of mass} \\ \text{flux of species} \\ \text{A into the} \\ \text{volume} \end{pmatrix} = (\rho_a u \epsilon b \, \Delta y)|_i + (\rho_a v \epsilon b \, \Delta x)|_j + (J_{ax} \epsilon b \, \Delta y)|_i \\ + (J_{ay} \epsilon b \, \Delta x)|_j + \rho_a' R \, \Delta x \, \Delta y \quad (3.23)$$

where the first two terms account for convection in the x and y directions, respectively, the next two terms account for the x and the y dispersion, respectively, and the last term accounts for the recharge.

The outflow expression is quite similar to (3.23) except that pumpage is considered rather than recharge. Here if Q, the pumping rate, is positive, the concentration of A carried by this flow is ρ_a. When the pumping rate is negative, indicating injection, the concentration may be specified. The concentration in the pumpage will be denoted ρ_a'' and the efflux term is

$$\begin{pmatrix} \text{rate of mass} \\ \text{flux of species} \\ \text{A out of the} \\ \text{volume} \end{pmatrix} = (\rho_a u \epsilon b\, \Delta y)|_{i+1} + (\rho_a v \epsilon b\, \Delta x)|_{j+1} + (J_{ax} \epsilon b\, \Delta y)|_{i+1} \\ + (J_{ay} \epsilon b\, \Delta x)|_{j+1} - \rho_a'' Q\, \Delta x\, \Delta y \quad (3.24)$$

The final term in (3.21), the chemical-reaction term, is very simple in form here but is very difficult to correlate accurately to concentrations, soil type, and flow condition. This net rate of chemical production is

$$\begin{pmatrix} \text{net rate at} \\ \text{which A is} \\ \text{produced by} \\ \text{chemical} \\ \text{reaction} \end{pmatrix} = R_a \epsilon b\, \Delta x\, \Delta y \quad (3.25)$$

Equations (3.22) through (3.25) may be combined according to the recipe of (3.21) to give

$$\frac{\Delta(\rho_a \epsilon\, \Delta x\, \Delta y b)}{\Delta t} = (\rho_a u \epsilon b\, \Delta y)|_i - (\rho_a u \epsilon b\, \Delta y)|_{i+1} \\ + (\rho_a v \epsilon b\, \Delta x)|_j - (\rho_a v \epsilon b\, \Delta x)|_{j+1} \\ + (J_{ax} \epsilon b\, \Delta y)|_i - (J_{ax} \epsilon b\, \Delta y)|_{i+1} \\ + (J_{ay} \epsilon b\, \Delta x)|_j - (J_{ay} \epsilon b\, \Delta x)|_{j+1} \\ + \rho_a' R\, \Delta x\, \Delta y - \rho_a'' Q\, \Delta x\, \Delta y + R_a \epsilon\, \Delta x\, \Delta y b \quad (3.26)$$

For the volume being considered, Δx and Δy are constants, so equation (3.26) may be divided by these parameters to give

$$\frac{\Delta(\rho_a \epsilon b)}{\Delta t} = \frac{(\rho_a u \epsilon b)_i - (\rho_a u \epsilon b)|_{i+1}}{\Delta x} + \frac{(\rho_a v \epsilon b)|_j - (\rho_a v \epsilon b)|_{j+1}}{\Delta y} \\ + \frac{(J_{ax} \epsilon b)|_i - (J_{ax} \epsilon b)|_{i+1}}{\Delta x} + \frac{(J_{ay} \epsilon b)|_j - (J_{ay} \epsilon b)|_{j+1}}{\Delta y} \\ + \rho_a' R - \rho_a'' Q + R_a \epsilon b \quad (3.27)$$

Then, using the Δ notation of (3.2) and (3.3), we obtain

$$\frac{\Delta(\rho_a \epsilon b)}{\Delta t} = -\frac{\Delta(\rho_a u \epsilon b)}{\Delta x} - \frac{\Delta(\rho_a v \epsilon b)}{\Delta y} - \frac{\Delta(J_{ax} \epsilon b)}{\Delta x} - \frac{\Delta(J_{ay} \epsilon b)}{\Delta y} \\ + \rho_a' R - \rho_a'' Q + R_a \epsilon b \quad (3.28)$$

This equation is the fundamental form of the species-balance equation. To solve this equation for p_a, the parameters ϵ, b, u, and v (from the previous discussion), R, Q, p'_a for positive recharge, and p''_a for well injection must be specified. The functional dependence of R_a on the chemical constituents present varies for different species, and no general form is known. However, some form for R_a, perhaps developed from laboratory experiments, must be specified and inserted into the equation. Also the functional form for J_{ax} and J_{ay} must be specified. Although success with this specification has been mixed, and the question is still open as to how best to specify the dispersion, current practice is to use a form such as

$$J_{ax} = -\rho D \frac{\Delta(\rho_a/\rho)}{\Delta x} \tag{3.29a}$$

$$J_{ay} = -\rho D \frac{\Delta(\rho_a/\rho)}{\Delta y} \tag{3.29b}$$

where D is the dispersivity and typically depends on velocity. More complex forms that account for anisotropic dispersion (the reasoning is similar to that in the discussion of anisotropic flow previously) are also used, but for simplicity we will use (3.29). The term ρ_a/ρ is a dimensionless quantity that is the mass fraction of A in the fluid. Recall that J_a is useful to describe mixing; and if ρ_a/ρ is constant, indicating a well-mixed system, $\Delta(\rho_a/\rho)$ will equal zero and there will be no dispersion. Substitution of (3.29) into (3.28) yields

$$\frac{\Delta(\rho_a \epsilon b)}{\Delta t} = -\frac{\Delta(\rho_a u \epsilon b)}{\Delta x} - \frac{\Delta(\rho_a v \epsilon b)}{\Delta y} + \frac{\Delta}{\Delta x}\left[\rho \epsilon b D \frac{\Delta(\rho_a/\rho)}{\Delta x}\right]$$
$$+ \frac{\Delta}{\Delta y}\left[\rho \epsilon b D \frac{\Delta(\rho_a/\rho)}{\Delta y}\right] + p'_a R - p''_a Q + R_a \epsilon b \tag{3.30}$$

When boundary conditions that specify the concentration or the flux of A at the boundary are used to supplement (3.30), this equation may be used to obtain the distribution of A throughout the study domain. Although equation (3.30) may appear to be complex, it states a very simple principle that per unit area, the rate of change of mass of A in a volume is equal to the net mass influx due to convection and dispersion plus the influx due to recharge, minus the efflux due to well pumping, plus the rate of production of A by chemical reaction. In the next section a very simple example will be worked that will give an indication of how flow equation (3.20) can be used in a numerical model. The procedure that would be adopted for equation (3.30) is similar, so no example will be presented.

3.10. EXAMPLE OF APPLICATION OF GROUNDWATER FLOW EQUATION

A computer model applies equations (3.20) and (3.30) to a groundwater aquifer that has been subdivided into hundreds or even thousands of small volumes. The computer is important to this model, because it can quickly perform the large number of arithmetic operations required to obtain the head or concentration distribution. The computer possesses no inherent wisdom; it can perform the necessary manipulations only if they are prescribed by the programmer. Thus the quality of computer results depends directly on the quality of the information and commands provided to the computer.

To give the reader a very simplistic view of how equations (3.20) and (3.30) are solved by the computer, we will apply equation (3.20) to the hypothetical confined aquifer depicted areally in Figure 3.4. The aquifer extends 5,000 meters in the x direction and 2,000 meters in the y direction. The aquifer is bounded below by an impermeable formation (aquiclude) but above by an aquitard, which recharges the aquifer uniformly at the rate of 50 cm/yr with water that contains no contaminants. The transmissivity, T, equals 3,000 m²/day. No flow crosses the northern, southern, or eastern edges of the aquifer. At the left edge, the hydraulic head ϕ is equal to 3 m above mean sea level. The aquifer is subdivided such that equations may be applied to the two square volumes in Figure 3.4.

Because the system is operating at steady state (i.e., conditions at any point in the aquifer do not vary with time), the time-difference term in equation (3.20) will be equal to zero. Further, there are no pumping wells in the system, so (3.20) becomes

$$0 = \frac{\Delta}{\Delta x}\left(T\frac{\Delta \phi}{\Delta x}\right) + \frac{\Delta}{\Delta y}\left(T\frac{\Delta \phi}{\Delta y}\right) + R \qquad (3.31)$$

This equation is applied to volumes 1 and 2 with reference to Figure 3.5. In Figure 3.5 values of $\Delta\phi/\Delta x$ at the east and west boundaries of the volume are indicated, and because $q_y = 0$ on the north and south boundaries, $\Delta\phi/\Delta y$ must be zero on these ends. Note that $\Delta x = 2,000$ m. For volume 1,

$$\frac{\Delta}{\Delta x}\left(T\frac{\Delta \phi}{\Delta x}\right) = \frac{\left(T\dfrac{\phi_2 - \phi_1}{2,000 \text{ m}}\right) - \left(T\dfrac{\phi_1 - 3 \text{ m}}{2,000 \text{ m}}\right)}{2,000 \text{ m}}$$

$$= \frac{3,000 \text{ m}^2}{\text{day}}\left[\frac{\phi_2 - 2\phi_1 + 3 \text{ m}}{(2,000 \text{ m})^2}\right] \qquad (3.32a)$$

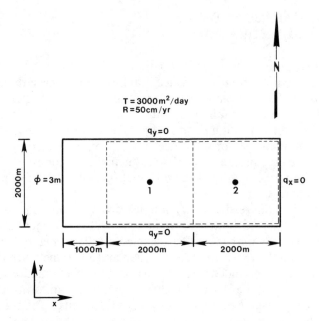

Figure 3.4.
Areal view of a hypothetical rectangular confined aquifer discretized into two elements.

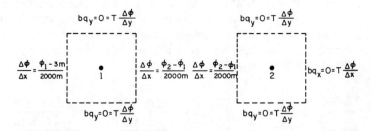

Figure 3.5.
Application of groundwater flow equation (3.31) to the two elements of the horizontal aquifer.

NUMERICAL SIMULATION OF GROUNDWATER CONTAMINATION

$$\frac{\Delta}{\Delta y}\left(T\frac{\Delta\phi}{\Delta y}\right) = \frac{0-0}{2{,}000\text{ m}} = 0 \quad (3.32\text{b})$$

$$R = 50\text{ cm/yr} = .00137\text{ m/day} \quad (3.32\text{c})$$

Insertion of (3.32) into (3.31) yields

$$0 = \frac{3{,}000}{(2{,}000)^2}(\phi_2 - 2\phi_1 + 3\text{ m}) + .00137\text{ m} \quad (3.33)$$

This single equation contains two unknowns and cannot be solved without a supplementary condition. The needed extra condition is provided by the balance over volume 2:

$$\frac{\Delta}{\Delta x}\left(T\frac{\Delta\phi}{\Delta x}\right) = \frac{0 - T\dfrac{\phi_2 - \phi_1}{2{,}000\text{ m}}}{2{,}000\text{ m}} = -\frac{3{,}000\text{ m}^2}{\text{day}}\left[\frac{\phi_2 - \phi_1}{(2{,}000\text{ m})^2}\right] \quad (3.34\text{a})$$

$$\frac{\Delta}{\Delta y}\left(T\frac{\Delta\phi}{\Delta y}\right) = \frac{0-0}{2{,}000\text{ m}} = 0 \quad (3.34\text{b})$$

$$R = 50\text{ cm/yr} = .00137\text{ m/day} \quad (3.34\text{c})$$

Combination of (3.34) into the balance equation (3.31) yields

$$0 = -\frac{3{,}000}{(2{,}000)^2}(\phi_2 - \phi_1) + .00137\text{ m} \quad (3.35)$$

Now equations (3.35) and (3.33) are two simple algebraic equations in the two unknowns ϕ_2 and ϕ_1, which we may solve simultaneously to obtain

$$\phi_1 = 6.65\text{ m}, \quad \phi_2 = 8.48\text{ m}$$

Because ϕ_2 is greater than ϕ_1, which in turn is greater than the western boundary value of 3.0 m, we can see that flow will move toward the west.

Once the solution for ϕ has been obtained, Darcy's law may be applied to compute the velocity field. This velocity field is used in equation (3.30) for computation of the concentration distribution. The differencing of this equation follows along the same lines as with the flow equation.

3.11. SUMMARY

This chapter has introduced the concepts behind numerical modeling. Numerical modeling is in no way magical but depends upon basic principles such as the conservation of mass. Although a number of techniques (e.g., finite difference, finite element) exist that are useful for modeling, they all function by discretization of the model region into subdomains and simultaneous application of the conservation principles to the subdomains. Errors in numerical models are due to inadequate data bases and, at times, to misapplication of models. Models cannot be used to generate data that do not exist; rather models are tools, based on existing data, that are useful in defining additional data needs and analyzing the expected behavior of a system based on the existing data.

CHAPTER THREE REFERENCES

BEAR, J., *Hydraulics of Groundwater,* McGraw-Hill, New York, 1979, 567 pp.

CHAPTER FOUR | MONITORING OF HAZARDOUS WASTE SITES

4.1. INTRODUCTION

The Resource Conservation and Recovery Act (RCRA) of 1976 proposes, when fully implemented, to provide "cradle-to-grave" regulation of hazardous wastes. Regulations under Section 265.F are concerned with groundwater monitoring. Groundwater monitoring requirements are also found in the Underground Injection Control program of the Safe Drinking Water Act and in several portions of the Surface Mining Control and Reclamation Act. Industry has gained experience in monitoring surface water and effluent as a result of the requirements of the National Pollutant Discharge Elimination Systems and the Federal Water Quality Act of 1965.

Effective monitoring of a hazardous waste disposal site is an extremely difficult data-collection problem. To understand its complexity, consider air pollution. Often we can see whether the pollution controls on a chimney are working; the smoke may be darkened and the odor (downwind) noxious. As the wind carries the polluted smoke, we can see and follow its direction.

Now imagine there are thousands of little chimneys around a factory site. By *looking* at the smoke, we may be able to tell which air pollution control devices are working and which are not. Again, we can see the trail of polluted smoke as it is carried away.

Imagine that we cannot see the sky, we cannot tell the direction or velocity of the wind, and we ask: Is the factory (with its thousands of little chimneys) polluting the air? That is our groundwater monitoring problem — at its easiest. It is made more difficult because the geological properties of the soil vary with depth and direction, and this variation is unknown or uncertain. When we look up in the sky, we observe the spatial variation of the pollutants. If we could look up only through a small tube or telescope, then the information we gathered from the one sighting might not be representative of what we would see if we looked everywhere. The small tube into the sky is like our groundwater monitoring well: the data we gather may not tell us too much about what is occurring in other nearby locations.

This chapter discusses approaches to the groundwater monitoring problem. It uses a conceptual framework to define various levels of monitoring and to distinguish between designing a monitoring network (made up of a number of observation wells) and operating the data-collection system over time. Some examples of monitoring will be presented to help the reader appreciate the effort required and the difficulties that may arise.

4.2. CONCEPTUAL FRAMEWORK FOR MONITORING

4.2.1. GENERAL MONITORING OBJECTIVES

Why should one want a groundwater monitoring system near hazardous waste sites? There are three primary reasons for developing such a system:

1. Assessment of existing or potential contamination from a site.
2. Acquisition of the information needed to develop or check a mathematical model of contaminant transport.
3. Design and evaluation of remedial actions for mitigating a particular contamination problem.

The second and third reasons go beyond the traditional concept that monitoring is concerned solely with contamination assessment. They address the need to understand the causes of contamination (i.e., develop models and test hypotheses) as well as the need to alleviate its adverse impacts (design remedial actions).

In most applications, the monitoring process can be divided into two related tasks:

1. The *design* task — specify the location of the wells and the parameters to be measured.
2. The *operation* task — specify how frequently parameters are to be measured.

These tasks are stated here simply, but the needs for the data change over time. This requires a dynamic design framework for the monitoring system. The monitoring effort should start with an initial design to assess the problem. After the problem has been clearly defined, more specific data may be required, particularly if a mathematical model is to be constructed or remedial measures developed. If possible, the data should not be collected all at once, but new well locations should be added incrementally as the need for additional information becomes apparent.

In order to clarify the amount of information required in different applications, it is convenient to define three levels of hazardous waste monitoring. Level 1 provides basic information for larger-scale regional planning. This information is generally used to identify whether or where specific problems exist. An example of Level 1 data is the report on organics in New York State drinking water (Kim and Stone, 1979). In this study, data on community water wells were compiled at the county level. Tables 4.1 and 4.2 illustrate the type of data collected.

Level 2 provides background data at a specific site. Such data could include information on groundwater flow paths and geology as well as a preliminary assessment of the extent of contamination. Level 3 provides the data required for detailed site evaluation and modeling or for the design of remedial actions at a specific site. The remainder of this section discusses the basic objectives of Level 2 and Level 3 monitoring networks. Methods for designing these networks are examined in Section 4.3. Some special considerations arising in groundwater monitoring are discussed in Sections 4.4 and 4.5. Important concepts introduced throughout the chapter are briefly reviewed in the final section.

The first step in setting up any monitoring system is to determine its objective. It is important to recognize that the particular objective selected has a significant impact on the design of the monitoring system and, in particular, on whether a Level 2 or Level 3 system is required. Some

Table 4.1 Organic Chemicals Detected in Community Water Supply Wells, Nassau County—4/28/78

Contaminant	Wells Tested	Wells Positive	Percent Positive	Maximum Level Detected ($\mu g/l$)
Tetrachloroethylene	372	57	15	375
1,1,2-Trichloroethylene	372	50	13	300
Chloroform	372	41	11	67
1,1,1-Trichloroethane	372	33	9	310
Carbon tetrachloride	372	20	5	21
Trifluorotrichloroethane	372	4	1	135

Table 4.2 The Ten Organic Chemicals Most Commonly Detected in Public Water System Wells — 10/78

Contaminant	Wells Tested	Wells Positive	Percent Positive	Maximum Level Detected ($\mu g/l$)
Bis(2-ethylhexyl)phthalate	39	36	92	170.0
Toluene	39	33	85	10.0
Di-n-butyl phthalate	39	21	54	470.0
Trichloroethylene	39	18	46	19.0
Ethylbenzene	39	17	44	40.0
Diethyl phthalate	39	13	33	4.6
Trichlorofluoromethane	39	11	28	13.0
Anthracene/phenanthrene	39	7	18	21.0
Benzene	39	6	15	9.6
Butyl benzyl phthalate	39	5	13	38.0

possible objectives, and their implications for monitoring and modeling, are discussed briefly in the paragraphs that follow.

Early Warning of Leachate Contamination: This objective is of primary interest to monitoring systems designed to fulfill the Resource Conservation and Recovery Act (RCRA) regulations. Early warning networks imply a Level 2 effort, where background geologic data is used to determine where leachates from a storage facility or landfill site are likely to migrate. This information is often insufficient to determine the rate of migration or to model its movement.

Determining the Rate of Leachate Plume Movement: In order to determine the rate and to simulate the movement of a three-dimensional contaminant plume in a groundwater system, it is necessary to account for all relevant transport processes. The processes that should be considered are convection, dispersion, and chemical reaction. *Convection* is the net flow due to gradients in the hydraulic head. *Dispersion* refers to mixing and spreading due to molecular motion and to deviation of the fluid velocity in the pores from the mean value. *Chemical reaction* is herein used in a broad sense that includes conversion of one chemical species to another as well as adsorption and desorption of the various contaminants on the solid matrix. Data collection at a finer scale than a Level 2 network is required if these transport processes are to be examined in detail.

The objective of determining the rate of leachate plume movement is closely related to the design of remedial schemes. One of the best ways to evaluate the design of a remedial measure is to construct a mathematical model of leachate movement that can be used to simulate the consequences of various pumping, trenching, and recharge alternatives. A model with such capabilities requires the support of a Level 3 monitoring network.

Ideally, this network should provide the data needed to estimate all unknown model inputs and parameters. It is particularly important to collect data over an extended period, since the model must be able to predict leachate movement when inputs such as rainfall, river and lake levels, and groundwater pumpage vary. Similarly, adequate spatial coverage is needed to enable the model to properly account for nonuniformities in site characteristics. When remedial actions are to be evaluated with a model, one should be prepared to carry out a reasonably extensive monitoring program; otherwise, there is no way to judge the reliability of the model's predictions.

Detection of Interaquifer Movement: The potential of a leachate plume to move into a previously unpolluted aquifer depends upon the water-pressure difference between the aquifers and the material properties of the soil separating the aquifers.

Information regarding relative water pressures can be determined by drilling two adjacent monitoring wells—one screened at the bottom of the top aquifer, the second screened at the top of the lower aquifer. Care must be taken to seal the annulus between the casing and the well hole where the well hole penetrates the confining layer to prevent flow between the aquifers. If the head elevation for the bottom aquifer is higher, then no potential exists for its being polluted, at that point, from the upper aquifer, since the flow would move upward. If the head elevation in the bottom aquifer is lower, then downward flow is possible. In this case, Level 2 geological surveys can provide estimates of hydraulic conductivity and porosity that can be used to determine the potential rate of downward flow.

Level 2 information consists primarily of data collected from existing wells, surface water bodies, and seeps, from existing geologic maps and logs, from surface geophysical investigation, and from a limited number of strategically placed new wells. These new wells should provide the greatest possible amount of information about aquifer characteristics and regional groundwater movement as well as about possible interactions between the hazardous waste site and the regional groundwater system. The most promising locations for new wells are usually deduced from a review of the existing well and geophysical data and possibly from simple mathematical models. Data from the new wells often suggest areas for further model development. They may also indicate that there is a need for a more comprehensive Level 3 monitoring effort. If a Level 3 monitoring network is set up, additional wells will usually be required in order to meet the more exacting modeling or remedial action design objectives associated with a Level 3 investigation.

The slow movement of groundwater makes it desirable to develop an iterative decision framework. In this framework information gathered at a few wells is used to update our understanding of the problem and reduce uncertainties. If the new information does not reduce our uncertainty

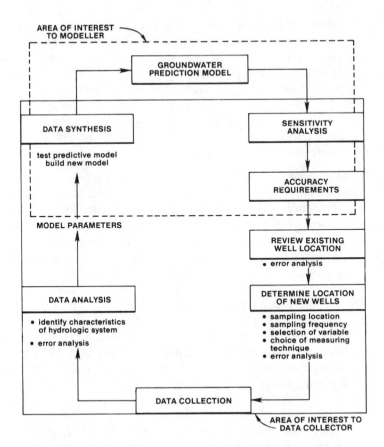

Figure 4.1.
Elements of the monitoring design process (after Moody, 1970).

sufficiently, then additional wells are drilled and the resulting information is added to the total data set. The decision process is then repeated. This iterative decision analysis framework has proven to be very effective in setting up both Level 2 and Level 3 networks.

The decisions that must be made at each iteration of the monitoring design process include the number of additional wells and their locations, the depth of each well, the screen length, the water-quality variables to be sampled, and the temporal sampling frequency. If a simulation model is used as part of the design process, further decisions must be made about the need for modifications in the model's structure or input parameters. The relationship between model-oriented and data-oriented decisions is shown in Figure 4.1, which portrays one investigator's view of the monitoring design process. This diagram indicates that sensitivity and accuracy information provided by the model can help the data collector decide where new wells should be located. Similarly, data collected in the field can help the modeler improve his understanding of the groundwater system he is simulating. It should be apparent that both the modeler and the data collector can benefit from close cooperation in the monitoring design process.

It is important to note in Figure 4.1 the prominent role played by error analysis—an aspect of monitoring that often is overlooked or minimized.

Errors accumulate during each step of the monitoring process: selection of well locations, well construction, sampling, laboratory analysis, data interpretation, and aquifer parameter estimation. It is important to analyze these errors so that the reliability of the data can be properly evaluated. Section 4.3.3 discusses in more detail the uncertainty that may arise in a typical monitoring study.

The conceptual framework outlined above for hazardous waste site monitoring recognizes the importance of different monitoring objectives and emphasizes the advantages of an iterative design process. The next section considers in more detail the application of our conceptual framework to particular monitoring problems.

4.3. MONITORING NETWORK DESIGN

4.3.1. BASIC DATA FOR MONITORING DESIGN

As discussed in Section 4.2, the purpose of Level 2 network design is to obtain general information on a site's hydrogeology and water quality. The primary sources of this information include surficial geology, bedrock geology, and some groundwater information. Each of these is discussed in one of the following subsections.

Surficial Geology: The purpose of a surficial geology survey is to determine the areal extent and thickness of the various deposits under and adjacent to the hazardous waste site. This information can be determined initially from regional mappings performed by the State Geological Survey. Such mappings indicate the deposits that one would expect at a site (sands, shales, major aquifer systems)—recognizing, of course, that important local variations may exist that do not show up. Other existing data will often include geologic logs of existing wells or test borings near the site. These logs (especially for wells) are usually filed with the State Geological Survey. Permeabilities of the various strata can often be estimated from the material encountered during drilling. Pump tests (a well-known hydrogeologic test) are used to estimate transmissivity and hydraulic conductivity. In critical areas where data are unavailable, a few wells should be installed. During their installation they will provide important geological information; later on they can be utilized as groundwater monitoring wells.

Bedrock Geology: The purpose of the bedrock geology survey is similar to that of the surficial geology survey. The difference is that interest is focused upon the bedrock beneath a site to determine whether it will act as a barrier. The survey will try to determine the material qualities of the bedrock, including the extent of fracturing. It is also important to know whether diabase dikes exist in the bedrock. These dikes form when hot magma is forced up between layers (often shales) and cools. If cooling takes place quickly, the dike material will fracture extensively. These fractured dikes can act as conduits for leachate, with the result that contamination can travel great distances rather quickly. These dikes can often be located using geophysical instrumentation such as a magnetometer.

Groundwater: The purpose of groundwater surveys in Level 2 networks is to determine regional flow patterns (both horizontal and vertical) so that possible interactions between the hazardous waste site and the groundwater system can be identified. The extent of a groundwater survey will vary, depending on whether the site is a landfill or an industrial storage site. In the former case, the site will have a greater influence upon the groundwater, and the survey will have to determine the extent of groundwater mounding caused by the (existing) landfill, the degree of influence the landfill has on the rate and direction of flow, and the rate of site infiltration relative to the total groundwater flow. This last item can be estimated through a water-balance analysis of the landfill site.

Required groundwater information, common to both industrial and landfill sites, include depth to water table, location of recharge and discharge areas, types and hydrologic interconnections of aquifers, and rate of site infiltration. Initially this information may be obtained from groundwater wells in place. Difficulties can arise when wells are drilled to different depth elevations, in that the head difference may not accurately describe the groundwater flow due to a vertical head component. Where information does not exist, wells will have to be installed. Besides the information described above, the wells should also give information concerning vertical head gradients within and between aquifers that may be separated by a clay layer. This implies clustering of wells, with each one measuring the water elevation in a different aquifer.

The three sources of information discussed above will determine the potential for a landfill or storage site to contaminate the groundwater. From the information gathered during the hydrogeologic investigation, a preliminary (Level 2) monitoring program can be designed and implemented. Wells installed to collect basic design information can usually be used for monitoring. It is worth noting that if the objective of the monitoring system is the early detection of contaminating leachate, then a well that is screened its entire length is the most effective. This may be at odds with government regulations, but experience has shown that many leachate plumes are quite thin, and the possibility exists that the plume will pass above (or below) a short screened length.

An example of a Level 2 monitoring design based on surficial geology and groundwater information is described in the next subsection. It is useful to keep in mind, while reading this example, the objectives and scope of the overall monitoring investigation.

4.3.2. EXAMPLE 1: THE SOUTH BRUNSWICK LEVEL 2 NETWORK DESIGN PROBLEM

Chapter 5 describes the South Brunswick contamination problem. When 1,1,1-trichloroethane was discovered in a municipal well, the firm of Geraghty and Miller, Inc., was retained to define the extent and movement of the contaminated groundwater in the vicinity of Municipal Well SB11. The objectives of the investigation were: (1) define the subsurface lithology, (2) determine the directions and rates of the groundwater movement, and (3) define the extent of the contamination plumes. Their investigation led to the establishment of a Level 2 groundwater well network. Our purpose here is to give readers an indication of the extent of the monitoring design effort. More details on the case study are given in Section 5.3.

To help establish the surface lithology near the contaminated well, lithologic logs of 74 wells were analyzed. These wells included three municipal water supply wells, seven private domestic wells, four monitoring wells at Mid-Eastern Aluminum Company, 30 monitoring wells on the IBM Corporation site, and 30 additional wells drilled for Geraghty and Miller, Inc., in the areas where data were scarce. The study area, shown in Figure 4.2, is approximately 7,500 feet by 3,500 feet; most of the investigative activity took place in a much smaller area adjacent to Municipal Well SB11.

To establish the extent of contamination, over 160 wells were sampled. The aquifers, both upper and lower, are sand aquifers (Old Bridge sand in the upper unit, Farrington sand in the lower) and they are separated by Woodbridge clay, which disappears in some areas (see Figure 5.24). The vertical head gradients within each aquifer are negligible, even though a difference exists between them where they are separated by the Woodbridge clay layer. Thus, one can use all the wells in the upper unit to estimate groundwater elevations without worrying about how or where they are screened.

Several hundred water samples were collected at the various monitoring wells, and from those analyses the extent of the contaminant plume could be estimated. The Geraghty and Miller study concentrated on the upper aquifer unit, and only a few wells were installed into the lower aquifer. The regional groundwater flow pattern was estimated by measuring well elevation levels at 75 wells. From the groundwater profile and limited field data concerning permeability, flow rates were estimated to be .75 to 2 feet per day.

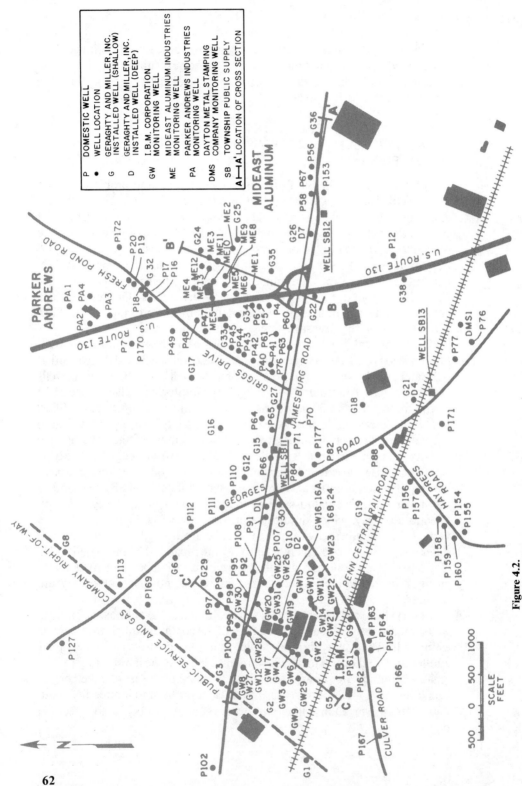

Figure 4.2.
Location of wells for the Level 2 analysis of Example 1—South Brunswick Township, New Jersey (from Geraghty and Miller, 1979).

The brief survey of the South Brunswick case study presented above indicates that a substantial amount of effort is involved with even a Level 2 network design. In this case study, a remedial scheme to clean up the aquifers had to be designed so a more detailed Level 3 monitoring network was required for the upper aquifer. The data from this network were used to estimate the inputs for a mathematical simulation model. The investigation concentrated on a .25 square mile area and over 55 wells were installed and water samples collected and analyzed weekly for eighteen months. This operation was expensive but was necessary to make the model's predictions credible.

4.3.3. DESIGN CONSIDERATIONS FOR LEVEL 3 MONITORING NETWORKS

A Level 3 monitoring network requires data collection at a scale appropriate for mathematical modeling, as shown in Figure 4.1. In setting up a Level 3 system, three basic questions arise:

1. How much effort should be expended on the network, and how should it be allocated to return the most information?

2. How many samples should be collected, and at what frequency?

3. What inferences can be drawn from data collected by the network, and what are the uncertainties (and confidences) in those inferences?

All three questions are closely interrelated, but to fully answer the first two, one needs to understand the uncertainty that can arise in monitoring data.

Most textbooks on groundwater indicate that variability in individual constituent concentrations in groundwater over time is typically small. The authors' experience, which is collaborated extensively by field data, is that variability can be extremely large where the groundwater has been affected by human activities. Sgambat and Stedinger (1981) give two situations that can lead to large variations:

1. Where wells are very near sources of contamination, so that fluctuations of pollutant inputs can lead to large variation in contaminant concentrations in the aquifer, and

2. Where wells are open at the top of the aquifer, so that short-term variation in the groundwater levels can flush pollutants out of the dry, unsaturated soil.

The variability of groundwater quality in wells adjacent to landfills, industrial sites, and other potential sources of contamination is of importance to monitoring. How much variability does one observe?

Sgambat et al. (1978) analyzed the variation of nitrate concentrations observed in eight shallow wells in Levittown, New York. The mean concentration was 9.6 mg/l and the standard deviation 6.4 mg/l. Sgambat and

Table 4.3 Analysis of Variations in Observed Nitrate Concentrations in Eight Wells in Levittown, New York[a]

Source of Variation	Contribution to Variance[b] $(mg/l)^2$	Percent of Contribution to the Variance
Measurement error	1	2
Within-well variation	11.7	29
Between-well variation	28.0	69

[a] After Sgambat and Stedinger (1981).
[b] Standard deviation equals the square root of the variance.

Stedinger (1981) analyzed the various sources of the observed variance, as shown in Table 4.3.

In cases of monitoring wells near landfills or industrial sites, the variability of concentration can also be large, as shown in Figures 4.3 to 4.6. Figure 4.3 shows the concentration of 1,1,1-trichloroethane in municipal well SB11 at South Brunswick, New Jersey (see Sections 4.3.2 and 5.3 for further discussion of this site). Figures 4.4 through 4.6 show the variability in three different organic pollutants in three wells adjacent to industrial sites.

Variability exists not only in sampling over time but in samples taken at essentially the same time. On December 21, 1977, six samples were taken from municipal well SB11 in South Brunswick, New Jersey. Table 4.4 presents the concentration of 1,1,1-trichloroethane. The highest concentration was almost six times the lowest.

Schmidt (1977) reported upon variation in measured nitrate concentrations as a function of time of pumping. Figure 4.7 presents the nitrate level from the time since pumping started. During the first three minutes of pumping the quality is variable and tends to be of high concentration. Over

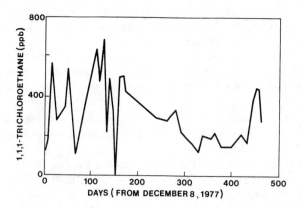

Figure 4.3.
1,1,1-trichloroethane concentration at well SB-11, South Brunswick, New Jersey (after Geraghty and Miller, 1979).

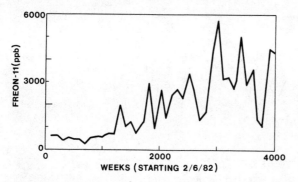

Figure 4.4.
Weekly freon-11 concentration.

the next 97 minutes there is a trend downward with a reduction in the variability.

Though the reason for it is not known, the downward trend has been widely observed in many field situations over a range of pollutants. Such observations point out the need for care in taking samples. The extraction and analysis of several samples at different times (say 5, 10, 20, and 40 minutes) would allow greater precision in estimating ambient concentration levels, although in many situations the low water yields of monitoring wells may make this impractical.

Since there are sources of uncertainty in any monitoring effort, it is reasonable to adopt, as a Level 3 network design criterion, the goal of minimizing uncertainty, subject to a specified constraint on network cost. In order to pursue this idea further, it is useful to consider the design of a monitoring network that provides information about the inputs to a groundwater simulation model. We will presume that this model is used to evaluate the relative effectiveness of alternative remedial measures at a

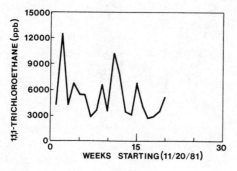

Figure 4.5.
Weekly 1,1,1-trichloroethane concentration.

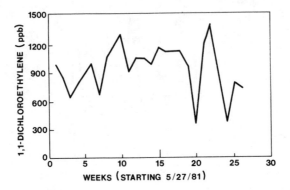

Figure 4.6.
Weekly 1,1-dichloroethylene concentration.

particular hazardous waste disposal site. Once the model's basic structure has been defined, the value of its independent variables (inputs) must be derived. The independent variables include infiltration and pumpage rates, hydraulic conductivity and soil porosity. The dependent variables (outputs) are usually groundwater elevations or groundwater velocities. The derivation of the independent-variable values requires a Level 3 field sampling or monitoring program.

The cost of field sampling programs that provide detailed information on all required model inputs and parameters is prohibitive for most studies. It is important, then, to design efficient monitoring systems that balance out the information obtained with the effort expended. Thus it follows that monitoring design variables such as well locations, sampling times, and number of replicate sample sets should be selected to maximize the model's prediction accuracy for a specified expenditure of sampling effort (Moore and McLaughlin, 1979).

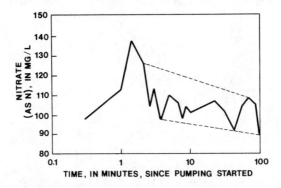

Figure 4.7.
Variations of measured nitrate concentrations in water from deep wells open to zones of differing water quality, as a function of time of pumping (after Schmidt, 1977).

Table 4.4 1,1,1-Trichloroethane Concentration for Six Samples Taken on 12/21/77 at Well SB11, South Brunswick, New Jersey[a]

Sample Number for 12/21/77	1,1,1-Trichloroethane (ppb)
1	180
2	570
3	1,060
4	220
5	660
6	680

[a] From Geraghty and Miller (1979).

There are a number of ways to estimate prediction accuracy, given certain assumptions about the relationship between model inputs and outputs as well as additional assumptions about the relationship between monitoring design variables (such as well location or sampling frequency) and the uncertainty of model inputs. Most of these methods for evaluating accuracy are based on statistical theory that is beyond the scope of our discussion. The types of results obtained are briefly summarized in the examples below.

4.3.4. EXAMPLE 2: DESIGN OF A LEVEL 3 MONITORING NETWORK FOR ESTIMATING MODEL HYDRAULIC CONDUCTIVITIES AND BOUNDARY CONDITIONS

This example, adapted from Moore and McLaughlin (1979), is concerned with determining the location of monitoring wells such that the independent parameters of particular groundwater model can be estimated. In this example, the steady-state groundwater elevation, due to steady inputs and boundary conditions, is given by a relatively simple equation:

$$y = -B(k)^{-1}D \tag{4.1}$$

where

y = water surface elevation evaluated at node points defined on a modeling grid (see Chapter 3),

$B(k)$ = matrix of parameters dependent upon the model structure and soil hydraulic parameters, k, and

D = set of model inputs (infiltration, pumpage) and boundary conditions, all assumed to be constant with time.

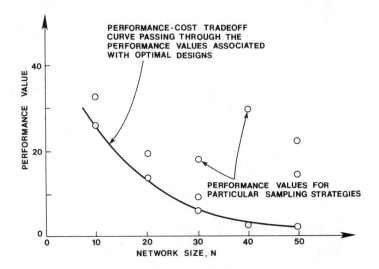

Figure 4.8.
Plot of a typical sampling design performance function versus number of samples for various sampling strategies (after Moore and McLaughlin, 1979).

The only unknowns are the soil hydraulic conductivity, k, and the boundary conditions in D.

Field monitoring results in N (uncertain) water elevation observations taken at N well locations, which we will specify as $x_1, x_2, \ldots, x_N$, distributed throughout the aquifer. As the location and the number vary, the accuracy in estimating $B(k)$ and D changes. Figure 4.8 presents the results of a typical sampling design performance analysis. The vertical axis is a performance index that is to be minimized (either sum of parameter estimation error or model forecast error). The horizontal axis is the total number of wells. For any set number of wells there are designs of varying efficiencies. The most efficient design is the one that minimizes the performance criterion for a set network size. This is shown as the performance-cost tradeoff curve in Figure 4.8.

In this example the problem was formulated for convenience in terms of a fixed design based on a total number of wells. In practice, the overall sampling program will evolve sequentially.

4.3.5. EXAMPLE 3: EXPANSION OF AN EXISTING MONITORING NETWORK*

Assume that eighteen monitoring wells exist in a region and the expansion of the network from eighteen wells up to 22 is considered. Four alternatives are being analyzed—adding 1, 2, 3, or 4 wells.

---

* After McLaughlin, 1981.

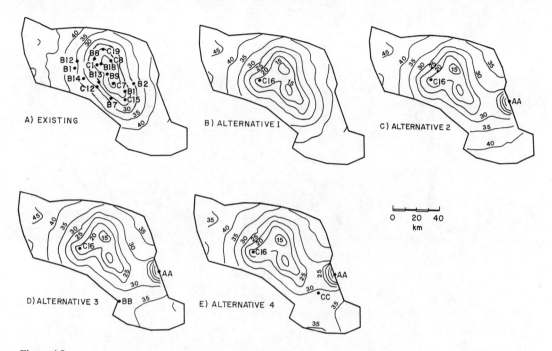

Figure 4.9.
Contours of error standard deviations for the existing network and alternatives 1–4 (after McLaughlin, 1981).

Figure 4.9(a) presents the network and contours of error standard deviations in the water-surface elevations. Changes in the error standard deviations can be computed. These are shown for each alternative in Figures 4.9(b) to 4.9(e). As expected, the standard deviation of the error gets reduced.

An estimate of the gain in information can be made in a variety of manners. Table 4.5 gives one such measure — the gain in information being measured as the reduction in the standard deviation of the estimation error. Assuming that the cost for each new additional is equal, a curve of information gain versus cost can be made, as shown in Figure 4.10.

Table 4.5 Average Error Standard Deviations and Information Return Measures for Borehole Sampling Alternatives 1 through 4

Alternative	Prior Standard Deviation (m)	Posterior Standard Deviation (m)	Information Return (m)
1	36.00	35.44	.66
2	36.00	32.22	3.78
3	36.00	30.33	5.67
4	36.00	28.56	7.44

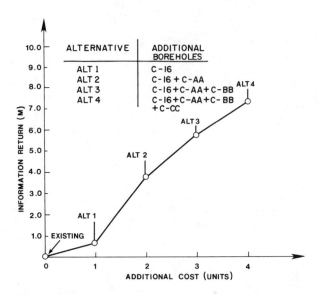

Figure 4.10.
Information return for the four alternatives of Example 3.

4.4. SURFACE GEOPHYSICAL MONITORING TECHNIQUES

A Level 2 investigation requires information on site geology, water-table location, and regional groundwater movement. If the available information is insufficient, then the use of surface geophysical techniques can be effective in collecting the data.

Geophysical methods have been effective in collecting two types of information: (1) analysis of the geologic stratigraphy, which includes boundaries between rock types and water-table depth variation, and (2) areal measurement of a leachate plume.

The advantage of surface geophysical techniques is that they provide continuous information over the site. Wells provide detailed information only at a point. Geophysical methods are most effective when combined with information gathered from a few wells. This is because much of the geophysical data can be interpreted only qualitatively unless it is related to soil information gathered from drilling.

4.4.1. DETERMINING STRATIGRAPHIC BOUNDARIES

A *stratigraphic boundary*—the boundary between two rock or soil types— is important, because such boundaries significantly influence the groundwater flow system. Two approaches to determining stratigraphic boundaries are seismic methods and ground-penetrating radar.

Seismic methods, as employed in groundwater studies, measure the refraction of energy waves from the interface between soil or rock bodies. The energy source is often a small explosive charge. The energy waves move through the ground and are sensed by a detector. The time lapse between the sending of the energy waves and their arrival at the various detectors allows for a plot of time lapse versus distance to the detector. From this plot, the depth to each soil/rock horizon can be estimated.

Owing to the complex nature of geological systems, great experience is often required in fully interpreting seismic information. Nevertheless, seismic refraction methods are effective in mapping a bedrock surface, determining water-table elevation, and mapping stratigraphic information.

Ground-penetrating radar is another technique for determining boundary and stratigraphic information. A single unit sends out and receives the radar signal. While the technique is effective in mapping bedrock surfaces and changes in soil density, its effective penetration depth is only about four meters. Furthermore, high-density surface features (such as a clay cap on a landfill) will reflect the signal, preventing any mapping below them.

4.4.2. DETERMINING LEACHATE PLUMES

Two surface geophysical techniques have been used to determine the extent of leachate plumes: electrical resistivity and electromagnetic conductivity. *Resistivity,* or inverse conductivity, of a material measures the resistance of a material to either an electrical or an electromagnetic current.

Conductivity in groundwater systems is a function of the interconnections and pathways of the pores, the presence of fluid in the pores, and the conductivity of the fluid. If the conductivity of the fluid is changed by contaminants (increased contamination usually leads to increased conductivity), then this change can be measured and used to map the extent of the leachate plume. This method assumes that the geologic structure does not vary, since variation in pore sizes and pathways would also affect the conductivity mapping.

4.5. OPERATION OF A MONITORING NETWORK

The operation of monitoring networks is concerned with the frequency with which parameters are sampled. A complete water-quality analysis for organic contaminants can cost up to $2,000. Clearly, the number of samples should be kept to a minimum. Quantitative analysis of the sampling-frequency problem has been performed by Wood and Mehra (1980) for surface-water quality, but limited research has been carried out for the groundwater monitoring problem. It turns out that a mathematical prediction model can be combined with sampling data to reduce the frequency of sampling.

The monitoring network and the prediction model provide two uncertain estimates of the pollutant concentrations. The first estimate comes from the surveillance data and is in error for reasons of laboratory procedures, sample representativeness, and so forth, as discussed in Section 4.3.3. A second estimate is based upon the physical mathematical equations and is in error since they may not represent the real world exactly (see Chapter 3 on modeling).

What is desired is a procedure for taking these two pieces of imprecise information and combining them so as to produce more precise, joint information. Fortunately, the mathematical apparatus is available; Figure 4.11 presents a schematic diagram of its procedure. Imagine that $x_1(t)$ and $x_2(t)$ are pollutant concentrations at some time t; the notation $\hat{x}(t|t)$ should be read as "the concentration estimate at time t using the model and measurements at time t."

As predictions are made in time, uncertainty grows. When the uncertainty is too large, a monitoring sample is taken. Because the prediction

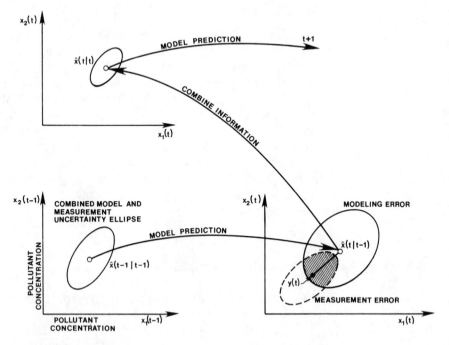

Figure 4.11.
Schematic of uncertainty propagation from measurement and model errors.

MONITORING OF HAZARDOUS WASTE SITES

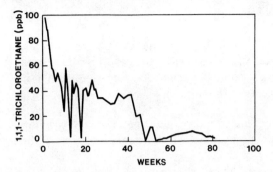

Figure 4.12.
Weekly concentration of 1,1,1-trichloroethane.

uncertainty depends upon the errors in the mathematical simulation model (evaluated during the model validation phase) and the surveillance errors (estimated from EPA *Standard Methods* and from the data analysis during the network design and model calibration phase), the frequency of sampling for long-term monitoring can be established.

This procedure was applied to the data shown in Figure 4.12, consisting of weekly measurements of 1,1,1-trichloroethane concentration for 81 weeks. The measured concentrations have a mean of 24.5 ppb with a standard deviation of 21.8 ppb.

Using the notation in Figure 4.11, the size of the combined model and measurement uncertainty is 5.3 ppb, which is about 25 percent of the uncertainty in the original data. This combined uncertainty grows during the model prediction step to 7.9 ppb (which is still only about 36 percent of the original uncertainty).

This uncertainty grows until a new measurement is taken. The sampling frequency was varied from one sample per week to one every four weeks to see how different frequencies affected the uncertainty in the concentration estimation. Table 4.6 presents these results.

Table 4.6 Standard Deviation in the Concentration Estimation Error (ppb) for Different Sampling Programs for Data in Figure 4.12.

Week	One Sample Every Week	One Sample Every Two Weeks	One Sample Every Three Weeks	One Sample Every Four Weeks
1	5.3	5.7	6.0	6.2
2	5.3	7.9	7.9	7.9
3		5.7	9.9	9.9
4			6.0	11.6
5				6.2

4.6. CLOSING COMMENTS

Groundwater monitoring systems are needed to evaluate the impact of hazardous waste disposal sites on the groundwater system. This chapter has put forth a conceptual framework for monitoring that recognizes two important facets of the problem:

1. Different information requirements lead to different network design levels. Three levels are proposed, of which Levels 2 and 3 are of direct relevance to this monograph.

2. Our approach to monitoring design is fundamentally statistical—especially for Level 3 networks. Our approach recognizes the importance of measurement variability and uncertainty and of the tradeoffs that exist between network size and accuracy. This statistical orientation suggests that the performance of a monitoring network is most appropriately measured by the accuracy of associated groundwater model predictions and parameter estimates.

CHAPTER FOUR REFERENCES

GERAGHTY AND MILLER, INC., "Investigation of Groundwater Contamination in South Brunswick Township, New Jersey," Syosset, NY, 1979.

KIM, N. K., and D. W. STONE, *Organic Chemicals and Drinking Water,* New York State Department of Health, Albany, 1979, 132 pp.

McLAUGHLIN, D., "Systematic Design of the BWIP Borehole Sampling Program: Program Report No. 3 on Task 4—Uncertainty Analysis," Rockwell Hanford Operations Energy Systems Group, Richland, WA., 1981.

MOODY, D. W., "Organizing a Data Collection Program for Water Resources Planning, Development and Management," *Intl. Symp. on Data Needs for Water Quality,* Univ. of Wisconsin, Madison, 1970, pp. 325–335.

MOORE, S. F., and D. McLAUGHLIN, "Hanford Groundwater Modeling—Review of Sampling Design Methods," *Report RHO-C-20,* Rockwell Hanford Operations Energy Systems Group, Richland, WA., 1979.

SCHMIDT, K. D., "Water Quality Variation for Pumping Wells," *Groundwater,* Vol. 15, No. 2 (1977).

SGAMBAT, J. P., and J. R. STEDINGER, "Confidence in Ground-water Monitoring," *Groundwater Monitoring Review,* Vol. 1, No. 1 (Spring 1981).

SGAMBAT, J. P., K. S. PORTER, and D. W. MILLER, "Regional Groundwater Quality Monitoring," AWRA: Symposium Proceedings of Establishment of Water Quality Monitoring Programs, June 12–14, 1978, Minneapolis.

WOOD, E. F., and R. F. MEHRA, "Model Identification and Sampling Rate Analysis for Forecasting the Quality of Plant Intake Water," IFAC, *Water and Related Land Resource Systems,* Cleveland, 1980, pp. 419–425.

CHAPTER FIVE | CASE STUDIES

5.1. A CASE STUDY OF LANDFILL LEACHATE GROUNDWATER CONTAMINATION

The Price Landfill in Pleasantville, New Jersey, was operated from 1960 through 1968 as a sand and gravel quarry. The landfill and vicinity are depicted in Figure 5.1. In 1967 the pit was excavated to within two feet of the water table. From then until November 1972, various forms of sanitary and industrial wastes were discarded at the site. Records indicate that between May of 1971 and November of 1972 approximately 9 million gallons of toxic and flammable chemical and liquid wastes were dumped at the landfill both in drums and through direct discharge onto the ground. Among the chemicals discharged were acetone, acid, chloroform, hexane, cesspool waste, oil, xylene, ethylene dichloride, and toluene.

The landfill is located approximately 3,400 feet west and upgradient of the nearest public supply well of the Atlantic City Municipal Utilities Authority (ACMUA). The ACMUA is charged with providing 10 to 15 million gallons per day of potable water needed by Atlantic City. Over 40 percent of this water has, in the past, been drawn from the upper Cohansey Sand, the geological formation most severely threatened by the pollution of Price's Landfill.

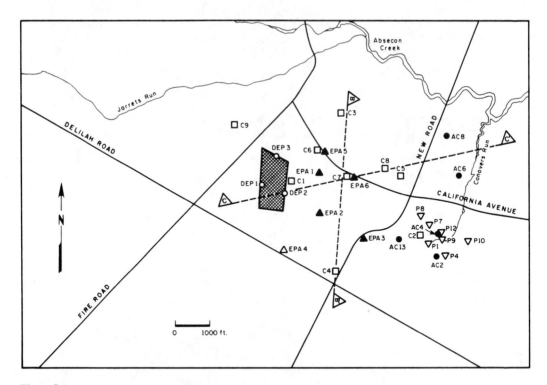

Figure 5.1.
Vicinity of Price Landfill (shaded region) with wells screened in upper zone of the Cohansey indicated (after Gray and Hoffman, 1983).

5.1.1. SITE INFORMATION AND CHARACTERISTICS

The Atlantic City region lies in the Coastal Plain Province of New Jersey and is underlain by unconsolidated sediments of Cretaceous to Recent origin. The formations dip gently to the southeast, outcropping in central New Jersey in bands running generally northeast-southwest.

Two major aquifers underlie Price's Landfill: the Cohansey Sand and the Kirkwood Formation, both of Tertiary age. The Kirkwood Formation consists of an 80-foot-thick water-producing sand unit approximately 800 feet below the land surface at Atlantic City, and an overlying clay bed 300 to 400 feet thick. The Cohansey Sand is subdivided into two water-producing units. The upper zone is approximately 80 feet thick, starting about 80 feet below the surface. It is underlain by a 50-foot-thick clay layer. The lower zone of the Cohansey Sand, the second water-producing unit, underlies the clay and is approximately 100 feet thick.

The Kirkwood Sand is relatively uniform and consistent, but the Co-

CASE STUDIES

hansey Sands are much more variable, with clay lenses distributed throughout. An indication of this variability is the fact that the clay layer separating the upper and lower zones of the Cohansey at the study site is not present in the Cohansey at other sites in New Jersey (Barksdale et al., 1936).

Over the Cohansey is an irregular clay unit of thickness varying from 0 to 30 feet. This clay is discontinuous and in places allows direct recharge to the Cohansey. Over the clay are sediments of Pleistocene and Recent age, which form a minor water-producing unit. Geologic cross sections BB' and CC' indicated on Figure 5.1 are depicted in Figures 5.2 and 5.3, respectively.

The locations of wells screened in the upper Cohansey in the vicinity of Price's Landfill are depicted in Figure 5.1. The Atlantic City (AC) wells to the east are the municipal watersupply wells that are threatened by the landfill leachate. Since 1973, twelve monitoring wells at nine different locations have been installed by the New Jersey Department of Environmental Protection (DEP) and the United States Environmental Protection Agency (EPA). The C and P series wells were installed in the summer of 1981 by the firm of Paulus, Sokolowski, and Sartor (PSS). The C wells (except for C8 and C9) are clusters of two or three separate wells screened at

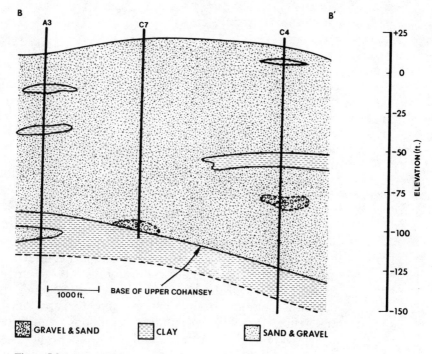

Figure 5.2.
Geologic cross section BB' of Figure 5.1: data from shallow and deep wells (after Gray and Hoffman, 1983).

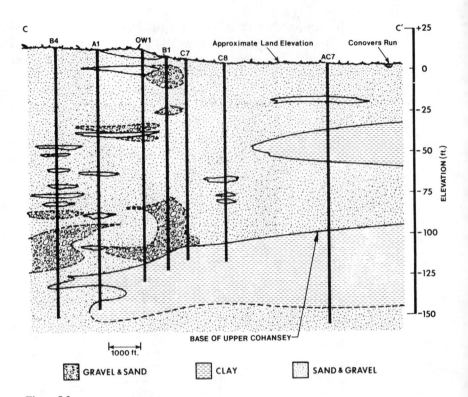

Figure 5.3.
Geologic cross section *CC'* of Figure 5.1: data from shallow and deep wells (after Gray and Hoffman, 1983).

different elevations in the upper Cohansey. They were installed to allow for detection of vertical flow gradients and possible stratification of the contamination plume. The P wells are single wells, installed to serve as observation wells for a pump test at well AC-4.

The available flow data were adequate to permit a two-dimensional study in the areal sense, wherein vertical gradients, which are small for this system, are neglected. The available pumping test data indicate that the aquifer responds to a change in pumping and settles into a new steady state in less than two days. For the time scale under consideration here, of the order of months and years, the effects of the short-duration transients are unimportant. Thus all simulations were run assuming that for a given pumping condition, steady-state flow conditions prevailed. It should be emphasized that different pumping conditions result in different steady-state head profiles. Under this assumption, the only flow feature that is not represented is the relatively rapid transition from one head configuration to another caused by changing the pumping strategy.

For two-dimensional flow, prime aquifer parameters for simulation of the head configuration are the storage coefficient and the transmissivity. The *storage coefficient* is important only for transient simulations, wherein it provides an indication of how quickly an aquifer will respond to a change in pumping. Therefore in this study it was not necessary to estimate the storage coefficient. *Transmissivity* is the volume of water transmitted per unit width of an aquifer over the whole thickness of the aquifer flow under a unit hydraulic gradient (Kruseman and DeRidder, 1979). Various sources report permeability values for the Cohansey ranging from 67 to 192 ft/day (Gray and Hoffman, 1983). Multiplication of permeability by the saturated thickness of flow provides the transmissivity. Actual pump tests conducted on the AC-4 well in 1981 by PSS revealed a transmissivity of 5,700 to 8,300 ft^2/day, which, for a saturated thickness of 100 feet for the Upper Cohansey in this vicinity, gives a permeability range of 57 to 83 ft/day.

Monthly pumping records of the ACMUA appear in Figure 5.4 for the period January 1969 through May 1981. The distribution of flow rate among the various wells is approximate, since ACMUA measures only the total flow. Head data at these wells prior to August 1981 was not available. However a set of head data for eleven of the DEP and EPA wells was collected in January 1981. A contour plot of these data is presented in Figure 5.5. The head data collected in the summer of 1981 were measured along section AA' of Figure 5.5 and appear in Figure 5.6. A similar gradient is noted for all cases, although the data demonstrate that at any given point

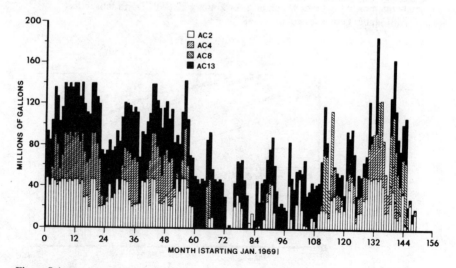

Figure 5.4.
Monthly pumpages from wells AC2, AC4, AC8, and AC13 in millions of gallons (after Gray and Hoffman, 1983).

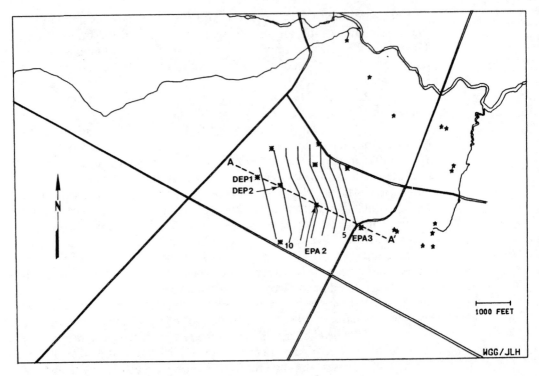

Figure 5.5.
Linearly interpolated water table contours for January 22, 1981 (elevations in feet above MSL) (after Gray and Hoffman, 1983).

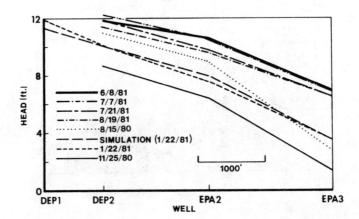

Figure 5.6.
Measured and simulated head gradient along section AA' of Figure 5.5 (after Gray and Hoffman, 1983).

or well, the value of the head may vary as much as two feet over time. The head data suggest that the direction of flow is in a direction slightly north of east—i.e., perpendicular to the lines of constant head.

Analysis of the piezometric data from the cluster wells revealed a common slight downward vertical gradient in the upper Cohansey of about .001 ft/ft. This gradient is small enough that the assumption of two-dimensional horizontal flow is not seriously violated. Examination of piezometric data for the Lower Cohansey indicates that the dividing clay layer within the Cohansey formation is capable of sustaining an 8- to 15-foot head difference across the formation, confirming that the interaction of the upper and lower zones is small.

5.1.2. MATHEMATICAL MODELING

A mathematical model of groundwater flow in the upper Cohansey aquifer was developed. Consistent with conclusions drawn in Section 5.1.1, the model assumes two-dimensional flow in the horizontal plane and steady-state conditions. The model equations, derived from the principles set forth in Chapter 3, were solved numerically using finite element techniques to give the head distribution in the vicinity of the study site. The region included in the model is covered by a triangular grid (Figure 5.7). The validity of the model was tested by comparing its output as shown in Figure 5.8 with the data of Figure 5.5. Pumping rates for the wells AC-2, AC-4, AC-8, and AC-13 were taken to be the January 1981 averages. Inspection shows that the magnitude and direction of the computed and measured head gradients agree quite well. Since the objective of this study was investigation of the long-term impacts of Price's Landfill, a simulation of the flow field using the ten-year average pumping rates for the abovementioned wells was also obtained (Figure 5.9). Note differences between Figures 5.8 and 5.9, which result from changes in pumping rate.

The mathematical model also included equations to simulate contaminant transport in the aquifer. This portion of the model was also two-dimensional in the horizontal plane but was different from the flow portion of the model in that it did not assume steady-state conditions. Therefore, the equations describing contaminant transport were transient; i.e., concentrations can change with time. The output from the flow model is used to describe the movement of contaminants in the groundwater and is therefore an input to the contaminant transport model. Contaminants were assumed to be conservative and nonadsorbing. The average water-table configuration of Figure 5.9 was assumed. Figures 5.10 and 5.11 depict the model predicted spread of the contaminant plume from Price's Landfill five and ten years, respectively, after the start of contaminant leaching to the groundwater. As is typically the case, a precise description of neither the contaminant leaching rate nor its temporal variation is known. Therefore a

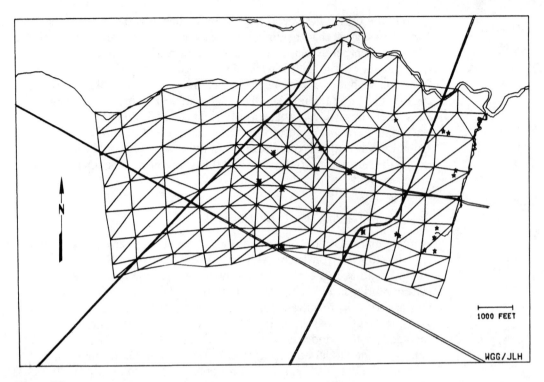

Figure 5.7.
Triangular finite element grid used in numerical simulation study (after Gray and Hoffman, 1983).

constant concentration along the eastern edge of the landfill was assumed for all time. The contour lines represent a change in concentration of 5 percent with the outermost line indicating a concentration 5 percent of that at the landfill site. Comparison with actual data was not possible, as such data were not available. Figures 5.10 and 5.11 present an indication of the direction and travel time of the contaminant and not a precise specification of the concentration and rate of change of that concentration throughout the study region.

5.1.3. EVALUATION OF REMEDIAL SCHEMES

The mathematical model was used to evaluate and compare the effectiveness of various pumping strategies that alter future migration of contaminants from the landfill. It was assumed that Price's Landfill would be "cleaned up" and no further contamination of the groundwater near the site would occur. The concentration at the landfill was therefore not held constant over time but allowed to decrease as the contaminant already in the

CASE STUDIES

aquifer was diluted by fresh water. The groundwater concentrations at the time of clean-up are assumed to be equal to those of Figure 5.11.

Two remedial schemes are discussed here. Remedial Scheme 1 (RS1) assumes that all current ACMUA pumping in the Upper Cohansey ceases. Figure 5.12 indicates the predicted water-table configuration if 10 million gallons per month are pumped from each of the locations near the ACMUA well field (indicated by "P" on the plot), decontaminated, and injected upgradient of Price's Landfill at the three locations indicated by "I". The philosophy behind this scheme is to flush all the pollutant from the system, making use of a gradient increased by the pumping and reinjection. The evolution of the initial concentration plume (of Figure 5.11) is shown after ten years of remedial pumping by Figure 5.13. It can be seen that the core of the contaminant plume is convected eastward and that the contaminant is diluted. Unfortunately the specified pumping is not strong enough to capture all the contamination, as indicated in Figure 5.13, where the edge of the plume has traveled past the pumping wells. Also the decrease in concentration over the ten-year period is disappointing, especially consider-

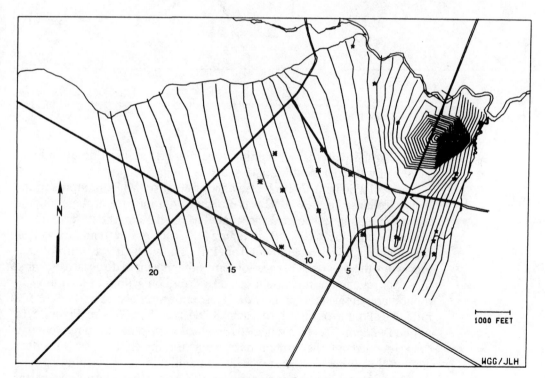

Figure 5.8.
Simulation of water table in the Upper Cohansey using average pumping rates for January 1981 (after Gray and Hoffman, 1983).

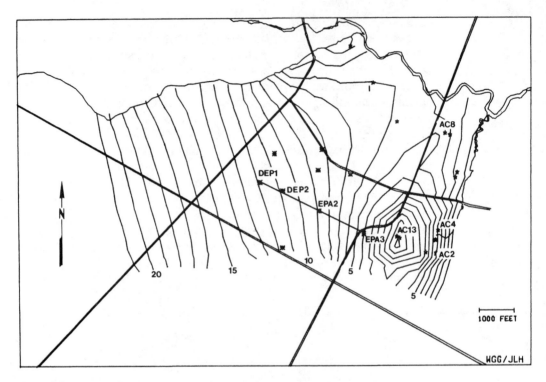

Figure 5.9.
Simulation of water table in the Upper Cohansey using average pumping rates of last ten years (after Gray and Hoffman, 1983).

ing the cost involved in pumping, treating, and reinjecting 30 million gallons of water per month for 10 years.

Remedial Scheme 2 (RS2) also assumes that ACMUA pumping in the Upper Cohansey ceases. RS2 was designed with the idea of removing the contamination at its source rather than flushing it out of the system. Three pumping wells, located by "P" in Figure 5.14, each pump 10 million gallons per month slightly downgradient from Price's Landfill. It is assumed that in processing the water and because of some other means of disposal (discharge to a river, sale to an industrial user) only 8 million gallons of the initial 30 million are reinjected each month. This reinjection occurs at the rate of 2 million gallons per month at each of the four locations in Figure 5.14 marked with an "I". This scheme is designed so that the pumping wells will draw in much of the contaminant while the injection wells will either contain the plume or at least provide some dilution at the leading edge.

The water-table contour plot appears in Figure 5.14. The initial concentration plume is that of Figure 5.11 and the effect of the remedial pumping is demonstrated after five, ten, and 25 years in Figures 5.15 through 5.17. During the first five years of pumping the zone of the highest concentration is convected slightly downgradient but the magnitude of the highest concentration is reduced significantly (Figure 5.15). At the same time the leading edge contour of the plume remains virtually stationary, so that the spreading of high-concentration pollutants is halted. During the second five years, the size of the contaminated zone decreases and further reduction of the maximum concentration is realized (see Figure 5.16). Finally, after 25 years (Figure 5.17) the contaminant plume is reduced as indicated.

Variations on this basic decontamination strategy may be investigated by examining the effect of altering pumping rates or of relocating some of the pumping wells after the first few years of operation. Possibilities of alternative remedial schemes are, of course, unlimited; but the inescapable fact is that the upper zone of the Cohansey aquifer can be cleaned only by a long-term and costly procedure.

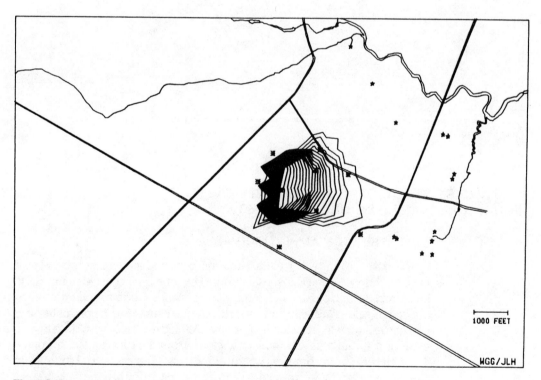

Figure 5.10.
Contaminant distribution after five years using the head distribution of Figure 5.9 (after Gray and Hoffman, 1983).

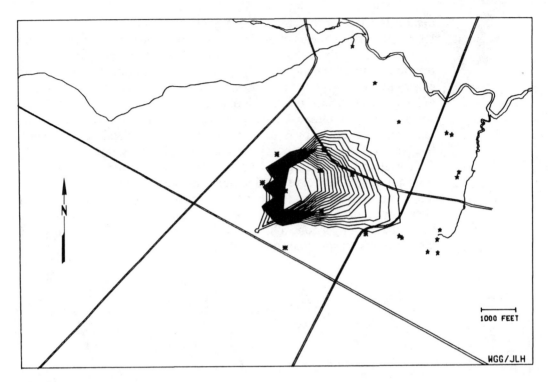

Figure 5.11.
Contaminant distribution after ten years using the head distribution of Figure 5.9 (after Gray and Hoffman, 1983).

5.2. A CASE STUDY OF A WASTE DISPOSAL POND CONTAMINATING GROUNDWATER

5.2.1. THE CONTAMINATION PROBLEM

In this case, groundwater contamination occurred as the result of seepage from an industrial waste disposal pond. The problem occurred in the South Farmingdale area of Nassau County, Long Island. Contamination was first noticed in June 1942. During a routine sanitary survey, 0.1 ppm of hexavalent chromium was detected in a private well located relatively close to a pit used for disposal of liquid industrial wastes from the Liberty Aircraft plant. The chief use of chromium was for anodizing aluminum and aluminum alloys to protect them from corrosion and to prepare their surfaces to take paints.

During World War II investigations of the problem were temporarily suspended. In 1945, when trained personnel again became available, a series of shallow test wells were installed in an area several hundred feet south of the aircraft plant. The samples collected from these wells indicated chromium concentrations from zero to a trace, and subsurface contamination was discounted as a threat to New York City's auxiliary groundwater system at Massapequa several miles to the south. The shallow test wells and an additional shallow domestic well about 1,500 feet south of the disposal pond were sampled in 1948, and 1.0–3.5 ppm of hexavalent chromium was reported along with cadmium, copper, and aluminum.

Recognizing the potential danger to public water supplies, the Nassau County departments of health and public works made a joint investigation of the contaminated area in 1949 and 1950. This program, which included the drilling and sampling of about 40 test wells, concentrated on defining the extent of contamination of toxic hexavalent chromium. By this time, about nine years after the start of disposal of the plating wastes, the contaminant plume had assumed an elliptical shape in plan view and was about 3,900 feet long with a maximum width of about 850 feet (Figure 5.18).

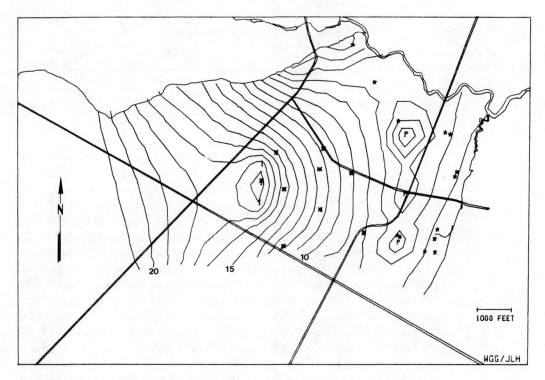

Figure 5.12.
Water-table contours for RS1 pumping 10 mgm from each well at "P" and injecting 10 mgm into each well at "I" (after Gray and Hoffman, 1983).

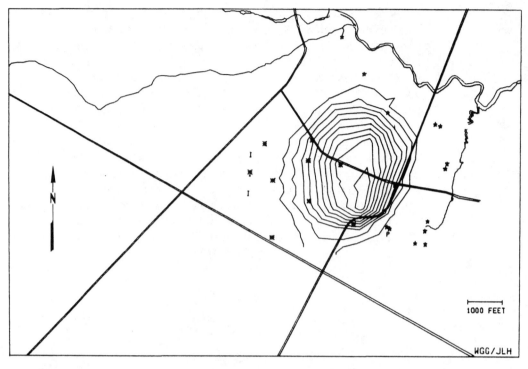

Figure 5.13.
Concentration distribution for RS1 after ten years (after Gray and Hoffman, 1983).

In 1953, 22 new sampling wells were constructed, and analyses were made for cadmium as well as chromium. Additional wells were drilled and samples collected by 1958 in the northern part of the plume; by this time the leading edge probably was a short distance west of Massapequa Creek.

In 1962, a detailed examination of the extent, chemical composition, and pattern of movement of the contaminated water was conducted by the U.S. Geological Survey in cooperation with the Nassau County departments of health and public works. About 100 test wells were installed and sampled between the plating-waste disposal basins and Massapequa Creek (Figure 5.18). To determine the vertical distribution of the chromium, samples were collected at 5-foot intervals until there was no detectable contamination. This investigation showed that the plume was about 4,300 feet long and had a maximum width of about 1,000 feet. The leading edge of the plume converged as it approached and moved beneath Massapequa Creek, where some of the contaminated water was discharged. The peak concentration had decreased from an observed high of 40 ppm in 1940 to 14 ppm in 1962. The maximum concentration at this time was located 3,000 feet down the direction of flow from the disposal ponds.

5.2.2. SUBSURFACE CHARACTERISTICS

The disposal ponds are located in an aquifer composed of glacial outwash material of Pleistocene age with a saturated thickness of 80 to 140 feet. The aquifer consists of beds and lenses of fine to coarse sand and gravel with thin lenses and beds of fine to medium sand and silt interbedded with the coarser material.

The glacial outwash aquifer is underlain by the Magothy aquifer, which includes the Magothy formation and younger undifferentiated formations of Late Cretaceous age. The Magothy aquifer is about 650 feet thick in the study area and is underlain by the Raritan clay, which acts as an aquitard. In some locations of the study area the Magothy aquifer is separated from the overlying glacial deposits by a silty sandy clay from 8 to 10 feet thick, but its continuity and lateral extent are unknown.

5.2.3. MATHEMATICAL MODELING

To investigate the physical mechanisms involved in the transport of hexavalent chromium at this site a mathematical model was developed and

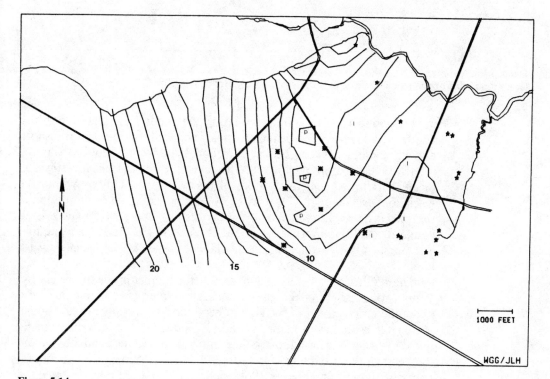

Figure 5.14.
Water-table contours for RS2 pumping 10 mgm from each well at "P" and injecting 2 mgm into each well at "I" (after Gray and Hoffman, 1983).

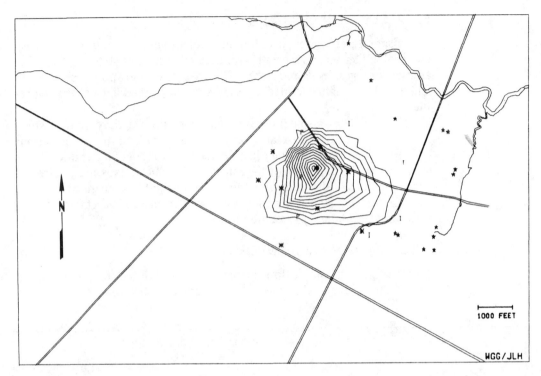

Figure 5.15.
Concentration distribution for RS2 after five years (after Gray and Hoffman, 1983).

applied (Pinder, 1973). In formulating the model, the glacial outwash aquifer was considered to be the principal vehicle of contaminant transport, and interaction between this aquifer and the Magothy aquifer was neglected. This assumption is justified for several reasons: (1) the investigation of 1962 indicated that the bottom of the plume was entirely within the glacial outwash aquifer, (2) a clay bed of unknown lateral extent underlies the glacial outwash aquifer in the area of study and probably acts as an aquitard, and (3) the vertical hydraulic conductivity of the Magothy aquifer is estimated to be less than 10 percent of that reported for the glacial outwash aquifer.

The aquifer parameters required to describe this problem are (1) hydraulic conductivity of the glacial outwash aquifer, (2) the thickness and hydraulic conductivity of the stream bed, and (3) the porosity. Although reasonable estimates of items (1) and (3) were available, there was inadequate hydrologic information to determine the hydraulic conductivity and thickness of the stream bed. In estimating these parameters, the contaminant distribution observed in 1962 played an important role. Because the leading edge of the plume was known to have moved beneath Massapequa

Creek and discharge of contaminated water into the creek was also evident, the hydraulic conductivity and thickness chosen for the stream bed must allow an accurate simulation of both of these observed phenomena. Through a trial-and-error procedure, a satisfactory simulation was achieved by using a hydraulic conductivity-to-thickness ratio of 10^{-3} per second.

Inasmuch as there had been no net change in the water-table elevations over the 1949–1963 period and no additional hydrologic stress was imposed on the system for this analysis, a steady-state groundwater flow pattern was assumed adequate for the transport simulation. The theoretical development of the mathematical model assumed that water density was a function of pressure only, that the chromium acted as a conservative ion, and that the removal of chromium by ion exchange during movement through this aquifer was negligible.

In simulating the movement of contaminants through the glacial outwash aquifer, it was assumed that chromium was present only at the disposal ponds and nowhere else in the aquifer. An extensive review of the literature revealed little information concerning the chromium concentration of the effluent residing in the ponds. Perlmutter and Lieber (1970) calculated a

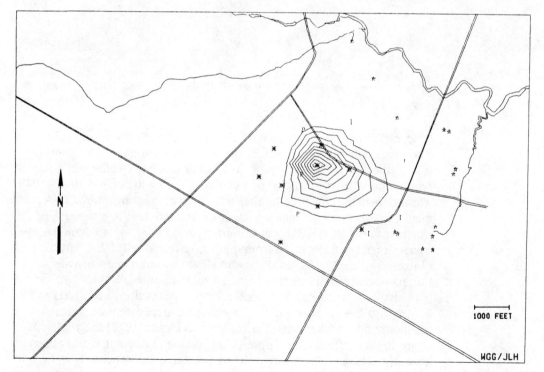

Figure 5.16.
Concentration distribution for RS2 after ten years (after Gray and Hoffman, 1983).

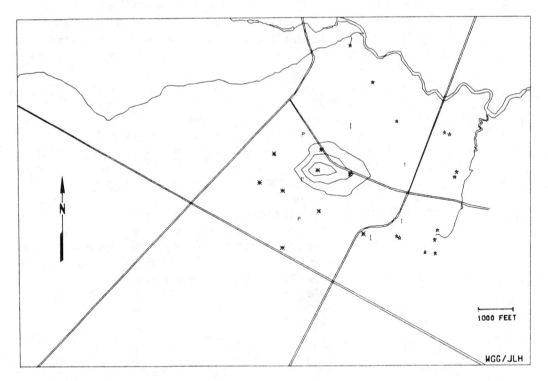

Figure 5.17.
Concentration distribution for RS2 after 25 years (after Gray and Hoffman, 1983).

concentration of 25 ppm based on chromic acid use per day and volume of discharged effluent. However, the observed maximum recorded concentration in the aquifer was 40 ppm. As a result this latter value was assumed representative of the disposal pond effluent during the period 1940–1949. The nature of the effluent after the installation of the chromium waste-treatment plant is open to question. Despite the reported effectiveness of the treatment process, it has not always been satisfactory, as was demonstrated by a series of examples taken from the disposal pond in 1962. Eight samples of treated effluent collected during an eleven-month period revealed chromium concentrations ranging from 3.1 to 35 ppm.

Calculated and observed contaminant distributions are presented in Figure 5.19 for the year 1953. The observed areal extent of plating-waste contamination is taken from Perlmutter and Lieber (1970) and represents a concentration of less than 1 ppm. Comparison of calculated and observed areal distributions shows a reasonably good match, particularly in light of the uncertainties inherent in the field data.

5.2.4. SENSITIVITY ANALYSIS AND MODEL PREDICTION

Once the mathematical model describing groundwater contamination was calibrated by using the historical data, several tests were conducted to learn more about the hydrologic aspects of the problem and the ultimate fate of the contaminant plume. Inasmuch as the chromium waste treatment plant installed in 1949 was designed to effectively remove all hexavalent chromium from the disposal pond effluent, the model was adjusted to simulate this situation. The computed configuration of the contaminant plume from 1949 to 1961 is indicated in Figure 5.20 and suggests that under these circumstances the contaminated groundwater should have essentially left the study area by the year 1961 either as discharge into the creek or as flow across the southern boundary of the model.

The chromium distribution actually observed during the period 1949–1962, however, is simulated not by eliminating all chromium from the disposal pond effluent but rather by reducing its concentration from 40

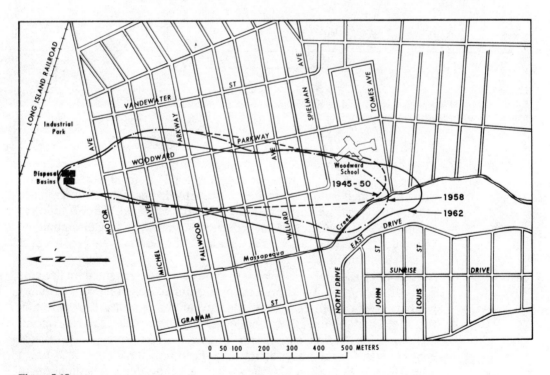

Figure 5.18.
Geographical extent of mathematical model and measured areal extent of contamination (after Pinder, 1973).

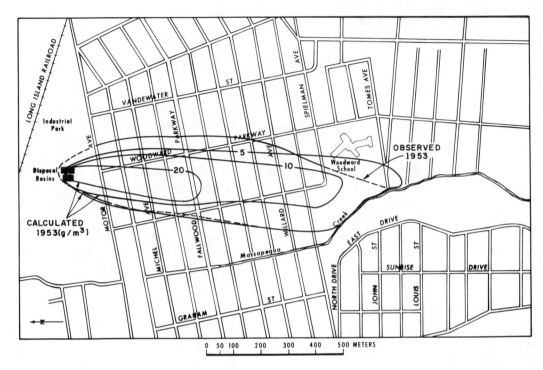

Figure 5.19.
Comparison of observed areal extent and calculated isopleths for chromium contamination in 1953 (after Pinder, 1973).

to 10 ppm. This specification is consistent with effluent analyses obtained during the sampling program conducted in 1962. If an effluent concentration of 10 ppm hexavalent chromium is assumed after 1949, a contaminant plume closely resembling the historical record is obtained. The evolution of the plume from 1949 to 1957 is indicated in Figure 5.21. Continuing the simulation after 1957, it became apparent after several years that a near steady-state situation had developed.

Inasmuch as groundwater velocities are generally very small (in this case 5.0×10^{-4} cm/sec), it was important to estimate how long contaminated groundwater would discharge into Massapequa Creek under the assumption that the disposal pond effluent no longer contained unacceptable concentrations of hexavalent chromium. Figure 5.22 illustrates the simulated evolution of the contaminant plume if a chromium-free effluent is assumed after 1972. The simulation shows that groundwater contamination of Massapequa Creek can be anticipated for about seven years after suitable remedial measures have been taken.

CASE STUDIES 95

5.3. A CASE STUDY OF ACCIDENTAL GROUNDWATER CONTAMINATION

5.3.1. THE CONTAMINATION PROBLEM

South Brunswick Township, New Jersey, utilizes groundwater for its public water supply. In December 1977 a common degreasing agent, 1,1,1-trichloroethane, was discovered in public supply well number SB11 (Figure 5.23). A private firm was selected by the township to determine the extent and probable cause of the contamination. The information presented here is taken primarily from Geraghty and Miller (1979).

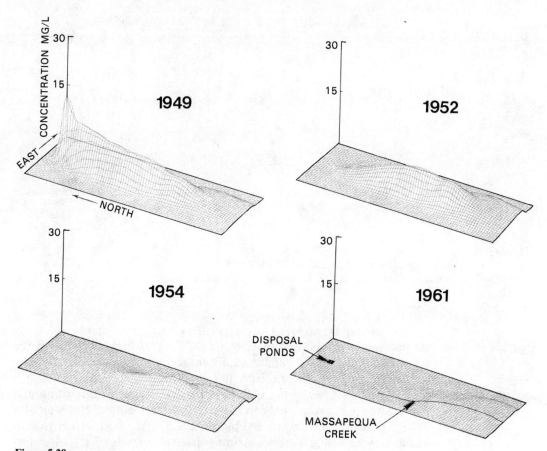

Figure 5.20.
Simulated areal distribution of contamination for period 1949–1951, assuming that the source of contamination is eliminated after 1949. Region represented appears in Figures 5.18 and 5.19 (after Pinder, 1973).

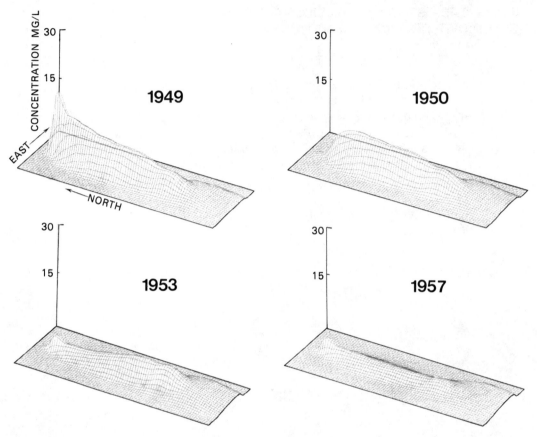

Figure 5.21.
Simulated areal distribution of contamination for period 1949–1957, assuming that strength of contaminating source decreases by 75 percent after 1949 (after Pinder, 1973).

The contaminant, 1,1,1-trichloroethane, has a solubility of 4.4×10^6 ppb (parts per billion) in water at 20°C. It is highly mobile in the subsurface and does not exhibit significant sorption on soil grains. Under newly proposed regulations, EPA has characterized this compound as a priority pollutant, while the U.S. Department of Health, Education and Welfare lists it as a suspected carcinogen. The acceptable level in water for short exposure, as recommended by EPA, is 33 ppb. It is further recommended by EPA that water containing detectable levels of 1,1,1-trichloroethane not be consumed over a long period. Other contaminants detected in well SB11 and local domestic and industrial observation wells include 1,1-dichloroethylene, trichloroethylene, tetrachloroethylene, benzene, toluene, zinc, and arsenic.

CASE STUDIES

5.3.2. SUBSURFACE CHARACTERISTICS

The township is located in the Coastal Plain physiographic province. It is underlain by 120 feet of unconsolidated sediment overlying shale. The unconsolidated sediment may be generally subdivided into a shallow and a deep aquifer, usually separated by a clay layer. The upper aquifer can be further subdivided into two units.

The upper aquifer consists of a clayey sand and gravel mix that readily absorbs and transmits precipitation to aquifers below. This overlying formation, called the Pensauken, provides small supplies of water to some wells for domestic or farm use.

The formation beneath the Pensauken in the upper aquifer is referred to

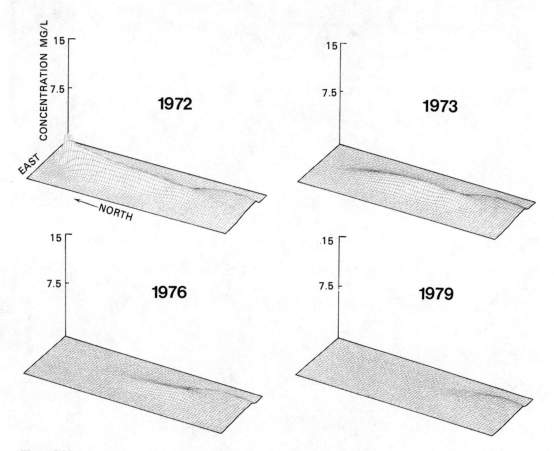

Figure 5.22.
Simulated areal distribution of contamination for period 1972 to 1979, assuming that the source of contamination is eliminated in 1972. Note change of scale for contamination concentration (after Pinder, 1973).

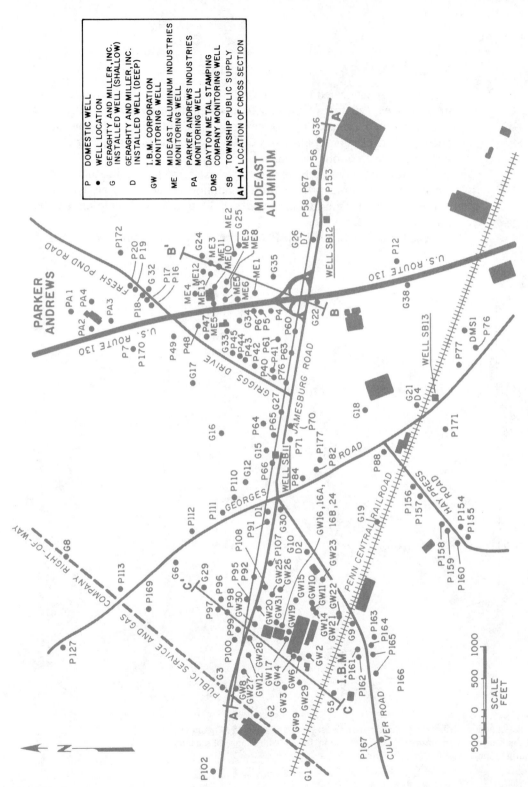

Figure 5.23.
Location of study area and wells used in this investigation—South Brunswick Township, New Jersey (after Geraghty and Miller, 1979).

CASE STUDIES

as the Old Bridge sand and frequently consists of fine to medium sand. This aquifer is quite productive and is a source for domestic wells.

The Old Bridge sand is separated from the lower water-bearing strata, the Farrington sand, by an impervious clay barrier referred to as the Woodbridge clay. This unit generally varies in thickness from 50 to 90 feet unless eroded away. The Farrington formation also consists of medium to fine sand, although its lower portion (10 to 20 feet thick) is a coarse sand with some localized gravel lenses. Water in this unit is generally suitable for most uses, for both industrial and municipal supplies.

Geraghty and Miller report that both the Old Bridge and Farrington aquifers are contaminated. The organic species most widely dispersed and usually appearing in highest concentration is 1,1,1-trichloroethane; hence it is used as an index of pollution. Geraghty and Miller propose that contaminated water is passing from the Old Bridge aquifer to the Farrington through an opening or "window" in the clay layer in the vicinity of well SB11, as can be seen in Figure 5.24. The cross-section locations correspond to Figure 5.23. The nature of the geologic substrata near well SB11 and depicted in Figure 5.24 is based on lithological data compiled from 30 holes drilled for the Geraghty and Miller study, 30 monitoring wells on the IBM Corporation site, four monitoring wells at an industrial site, public wells SB11, SB12, and SB13, and seven domestic wells. These wells are located on Figure 5.23.

5.3.3. THE ORIGIN AND EXTENT OF CONTAMINATION

An investigation was undertaken by Geraghty and Miller to establish those industries in the vicinity of well SB11 that could be potential sources of

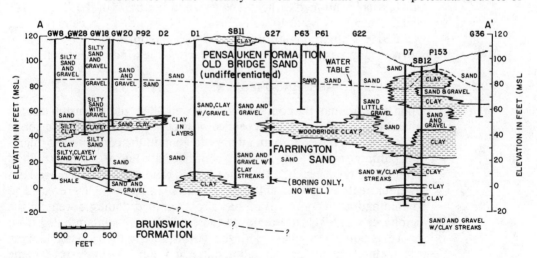

Figure 5.24.
Geologic substrata in vicinity of well SB11 (after Geraghty and Miller, 1979).

contamination. Figure 5.25 reflects the distribution of 1,1,1-trichloroethane in the area of investigation based on samples taken from over 160 wells. The contour lines describe the mean values of replicate analyses reported by various chemical laboratories. The contour line for 10 ppb, not shown, was nearly identical with that shown for 1 ppb. The information on Figure 5.25 indicates that the contamination of the upper aquifer has not resulted from a single source.

The movement of the contaminants can be understood only after the water table or potentiometric head distribution is established. A contour map of the water table in the area of interest, as of March 1979, is illustrated in Figure 5.26. This map would, of course, change seasonally, but it is probably representative of the flow pattern over a considerable part of the year. The direction of groundwater flow is along a line roughly orthogonal to the groundwater contours (material heterogeneity make this relationship approximate). The water-table and clay-window relationship is shown by section AA of Figure 5.24.

Using calculated groundwater flow rates and the observed contamination, Geraghty and Miller estimated that the IBM Corporation had been discharging organic chemicals from 1969 to 1978. Contaminants in the so-called IBM plume include 1,1,1-trichloroethane, tetrachloroethylene, 1,1-dichloroethylene, trichloroethylene, and zinc. Water originating as precipitation percolates through the ground at the IBM site and appears to be contaminated through contact with chemicals lodged in the unsaturated zone.

In the area under investigation, the lower Farrington aquifer is locally recharged by the Old Bridge aquifer via the breaks or discontinuities in the clay barrier mentioned earlier. The groundwater basin of the Old Bridge, which contributes water to wells SB11, SB12, and SB13, is centered near well SB11. Contamination flowing through the upper aquifer appears to move, in large part, into the lower unit through the window near well SB11.

5.3.4. REMEDIAL MEASURES

Upon realization of the contamination of well SB11, the Township discontinued its use and has since depended on wells SB12 and SB13 for public water supply. However, in an attempt to contain the contaminant migration, pumping of well SB11 has been continued at a rate of 0.4 mgd (million gallons per day). Since July 1978 the pumped water has been discharged to a sewer for treatment. Regular testing of wells SB12 and SB13 has revealed no contamination. The Township has also examined domestic wells in the vicinity of well SB11 and removed those which were contaminated from service as potable supplies. Both the Township and the EPA are looking into methods of treating the contaminated water. A successful scheme based on aeration is currently employed by the IBM Corporation in treating water pumped from beneath their site.

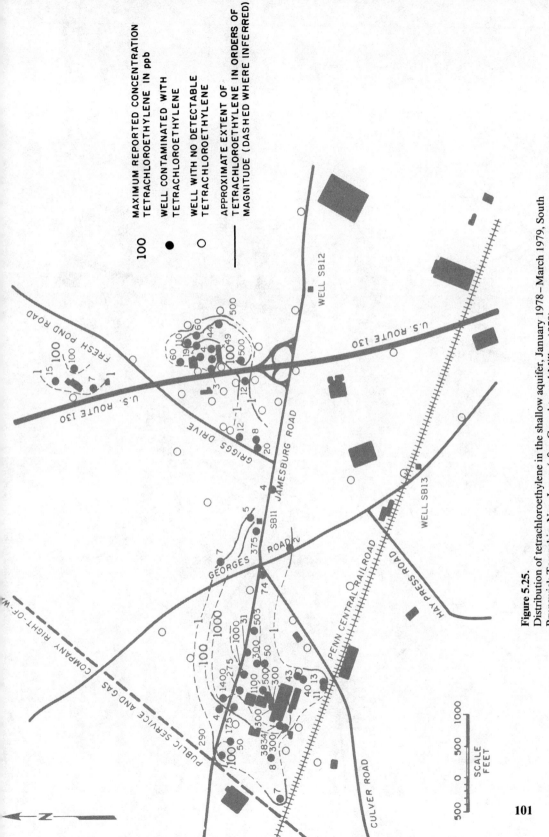

Figure 5.25.
Distribution of tetrachloroethylene in the shallow aquifer, January 1978–March 1979, South Brunswick Township, New Jersey (after Geraghty and Miller, 1979).

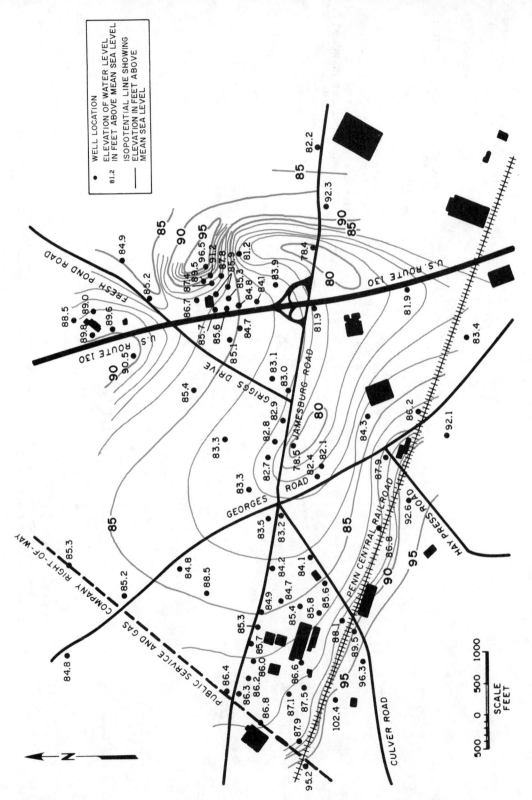

Figure 5.26.
Water-table configuration on March 9, 1979, South Brunswick Township, New Jersey (after Geraghty and Miller, 1979).

Through an agreement between the IBM Corporation, the State of New Jersey, and the Township of South Brunswick, IBM agreed to take action to renovate the upper aquifer between the IBM site and the window in the vicinity of well SB11.

The remedial program is a multistage project. In the first stage more field information was gathered. Additional wells were put in place to better define the lithology of the area and the water-level contours from which the direction of groundwater flow is determined.

The second stage involved the construction of a groundwater transport model. The model is based on a finite element numerical procedure. Contaminant transport is modeled in three dimensions on a time-variable basis. The model was calibrated against field information and adjusted until a satisfactory degree of confidence in the model forecasts was achieved.

The final stage of the project consisted of testing various remedial strategies using the calibrated model. A remedial strategy normally consisted of a series of withdrawal and injection wells strategically located on and around the IBM site. Approximately thirty different strategies were evaluated and the analyses presented to corporation management. Details of the analyses are not presented here, as they have not yet been approved for public dissemination.

Each analysis involved a simulation of the dynamics of the contaminant plume. The current geometry of the plume as estimated from model runs was used for the initial conditions. The contraction of the plume in response to the proposed strategy was then forecast by the model. The strategy selected by the corporation consisted of a series of discharge wells located about midway between the IBM site and the contaminated township well SB11. Recharge wells were located on-site between the assumed location of the source and the discharge wells. Construction of the remedial system is now complete and it is undergoing testing.

5.4. DISCUSSION OF CASE STUDIES

This chapter has presented three case studies of groundwater contamination by improper disposal or accidental spillage of hazardous materials on the land surface. While each instance is unique, they also exhibit certain common attributes. In the investigation and analysis of a groundwater pollution problem involving hazardous wastes, there is no established method or procedure. This reflects the fact that relatively few documented instances date back more than a few years. Nevertheless, one can propose an optimal course of action based on information to date.

The investigation is normally initiated through a chance observation of well contamination. The affected party normally begins by contacting a qualified groundwater hydrologist, who conducts a preliminary investiga-

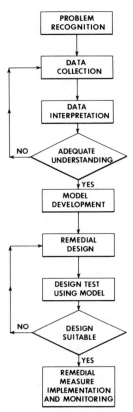

Figure 5.27.
Diagrammatic representation of contaminated groundwater investigation and restoration.

tion to establish the general scope of the problem. Inevitably, the existing data base is inadequate, and the hydrologist is asked to propose and perhaps oversee an initial data-collection program. In the cases cited above, this sequence of activities can be documented. There follows a period of analysis wherein one establishes the distribution and probable cause of the contamination. At this point it is often advantageous to consider the formulation of a mathematical model. While crude in design, owing to the limited data base, the model nevertheless pinpoints the most important data deficiencies. This initial analysis sets the stage for the long-term data-collection and monitoring program. Because of the importance of this stage, one often requests a second opinion from a competent specialist. The final stage is the design and implementation of the remedial scheme. At this stage some type of simulation procedure is virtually mandatory in order to forecast the effectiveness of proposed remedial schemes. Figure 5.27 presents an idealized schematic of the entire investigative and remedial program.

CHAPTER FIVE REFERENCES

BARKSDALE, H. C., R. W. SUNDSTROM, and M. S. BRUNSTEIN, *Supplementary Report on the Ground-Water Supplies of the Atlantic City Region,* State of New Jersey, State Water Policy Commission Special Report 6, 1936.

GERAGHTY and MILLER, INC., "Investigation of Groundwater Contamination in South Brunswick Township, New Jersey," Syosset, NY, 1979.

GRAY, W. G., and J. L. HOFFMAN, "A Numerical Model Study of Groundwater Contamination from Price's Landfill, New Jersey, I. Data Base and Flow Simulation," *Groundwater,* Vol. 21, No. 1 (1983), pp. 7–14.

———— and ————, "A Numerical Model Study of Groundwater Contamination from Price's Landfill, New Jersey, II. Sensitivity Analysis and Contaminant Plume Simulation," *Groundwater,* Vol. 21, No. 1 (1983), pp. 15–21.

KRUSEMAN, G. P., and N. A. DERIDDER, *Analysis and Evaluation of Pumping Test Data,* International Institute for Land Reclamation and Improvement, Bulletin 11, Wageningen, The Netherlands, 1979.

PERLMUTTER, N. M., and M. LIEBER, "Disposal of Plating Wastes and Sewage Contaminants in Ground Water and Surface Water, South Farmingdale–Massapequa Area, Nassau County, New York," Geological Survey Water Supply Paper 1879-G, 1970.

PINDER, G. F., "A Galerkin Finite Element Simulation of Groundwater Contamination on Long Island, New York," *Water Resources Research,* Vol. 9, No. 6 (1973), pp. 1657–1669.

CHAPTER SIX
DECISION FRAMEWORK FOR SITING HAZARDOUS WASTE FACILITIES

6.1. INTRODUCTION

Over the last few years the general public's awareness of the hazardous waste problem has greatly increased. Incidents such as Love Canal and the widespread publicity accorded those illegally dumping hazardous waste have served to frighten the public into a self-protective stance. In addition, the track record of hazardous waste disposal has not been good.

The U.S. Environmental Protection Agency (EPA) estimates that 90 percent of the hazardous waste generated in the United States has been disposed of by environmentally unsound methods and that 1,200 to 1,500 disposal sites present an imminent threat to human health (EPA, 1980). This poor track record coupled with the inherent risk and uncertainty in the disposal process has uniformly solidified public opposition to hazardous waste disposal facilities.

This opposition will have to be seriously reckoned with if any such facility is to get into operation. Thus, the siting of facilities will be an important disposal issue. While people can understand the need for disposal facilities on a national or regional level, no one wants one near his or her home or workplace, owing to the perceived danger. Wherever a facility is proposed to be sited, there will be strong local opposition. It is very important, therefore, that the best possible *method* be used in order to select the best sites; otherwise siting decisions will not stand up to fierce public opposition.

Before the characteristics of a good site-selection methodology can be developed, we need to clearly define the problem. The entire decision process required to get disposal facilities into operation is extremely complex. Many political, legal, and social factors interact on many levels.

The overall concern in the hazardous waste siting problem has been described by different sources to be the protection of water resources, public health, and the environment, or just general population well-being. These general concerns can be broken down into five nonredundant categories: the environment, economics, socioeconomics, health and safety, and public attitudes. This breakdown, suggested by Keeney (1980), defines the concerns on a manageable level, avoiding overly broad categories such as "human welfare," while at the same time not getting too specific.

The *environmental* concern deals with the effect the disposal facility has on the ecosystem of the site and its surrounding area owing to transportation of waste and to construction and operation of the facility. It does not include impacts on the human environment but the pollution of water, land, and air near the site. An example of an environmental concern would be the impact on aquatic life in a sensitive wetlands area due to the construction of a large disposal facility adjacent to it.

Economic concerns deal with site-specific costs incurred at a site due to its peculiar conditions. Costs, such as general construction costs, that are incurred to the same degree at every site are not included. The primary economic concern is to avoid any *extra* costs that are tied to building the facility in a specific location.

Socioeconomic concerns deal with the economic impact on people near the site caused by building a disposal facility. They include such factors as the benefit of additional taxes due to the facility, the devaluation of property due to the presence of the facility, and other such effects on the surrounding population.

Health and safety concerns are comparable to environmental concerns but deal with people. Primarily they deal with the increase in mortality, morbidity, and injury caused by normal operations and accidents at the facility. Here is where the impact of risk and uncertainty lies. An example of a health and safety concern would be the increase in cancer in the surrounding population caused by a disposal facility.

Public attitude concerns deal with the reaction of local citizens, environmental groups, industry, and other persons to a specific facility-siting decision. Even though two sites may be identical in the other four areas, the local reaction to them will vary with the makeup of the population.

A good methodology, then, will find the sites with the least adverse and most beneficial impact in these five multiobjective areas. With these assumptions and concerns in mind, we can now develop the characteristics of a good site-selection methodology.

6.2. THE DECISION PROCESS: FROM SITING CRITERIA TO SITES

A site-selection methodology is considered good if its outcome is selection of the best sites. This means two things. First, the sites selected must have the highest value or utility in the five concerns described above; that is, there must be no other sites that, by a similar evaluation, can be considered better with respect to all five concerns taken together. Second, the selections made must be justifiable; it must be clearly demonstrable how and why one site was chosen over another.

A good methodology must produce both these outcomes. Without the first there will be better sites available that are not chosen, and the methodology will have failed in its purpose to find them. The severity of this failure depends on the difference in quality between the sites. Since this is largely a matter of opinion, it is clear that the second outcome — the ability to justify site selections — is also quite important. Well-organized local opposition or special-interest groups may be able to defeat even the best site if it is not properly defended. Finding the best sites by way of a justifiable methodology, however, will give a chosen site the best chance of being used. This section will show the general characteristics of such a methodology and of the process by which it is worked out and sites are chosen.

6.3. THE OVERALL CHARACTERISTICS

A good site-selection methodology can be described by five characteristics: it should (1) be appropriate, (2) be formal, (3) include qualitative as well as quantitative considerations, (4) effectively involve the public, and (5) in the end be well explained. While these characteristics are by no means discrete, all the necessary characteristics will fit into one or more of them, and they do serve to explain the basics of a good methodology. If a methodology fails to include any one of them, it will, in theory, be lacking, and chances of finding the best sites and justifying the selection will be reduced. In reality, however, tradeoffs must be made, since no methodology will be able to fully include all these characteristics. The goal is to include each as much as possible. This section looks at the five categories in detail and shows why each is important.

6.3.1. APPROPRIATE

An *appropriate* methodology is one that is useful to the decision maker in making an informed siting decision. While this category does overlap the others — a methodology that is not formal, for example, will also be considered inappropriate — it does contain independent ideas.

For finding and defending sites for hazardous waste facilities, the methodology must reflect the magnitude and sensitivity of the siting problem. A methodology that is too shallow or *ad hoc* cannot deal with the complex issues involved. If the public perceives that sites have been chosen in an informal or careless manner, public opposition will make it impossible to use the sites. (The needs for a formal methodology and public involvement are dealt with below.)

6.3.2. FORMAL

The second characteristic of a good site-selection methodology is that it be *formal*. It should explicitly state data sources, assumptions, and value judgments and clearly indicate how the methodology was used to reach its conclusions. It should also state clearly what was not considered in the problem.

Such a formal analysis is important for three reasons. First, owing to the scope of the hazardous waste facility siting problem, it is difficult to address it, evaluate alternatives, and compare sites in an informal *ad hoc* manner. The problem is simply too large and complex to understand intuitively.

Second, a formal analysis is needed because no overall experts exist. The information needed to make an informed decision is far beyond what a single person can handle; rather, a team of experts from many different disciplines is needed. Furthermore, the problem involves many interest groups and other bodies, such as environmental commissions, whose opinions must be taken into account. To organize the efforts and inputs of so many people, a formal methodology is necessary.

Third, the decisions made will have to be justified. It would be difficult to explain a decision based on intuitive processes and unstated assumptions. With a formal methodology, however, everything is written down, and it becomes clear how a decision was reached and how the conclusions are justified.

A formal analysis, then, offers two major benefits:

1. The quality of the decision is greatly improved, because diverse groups of people with different qualifications and insights can look at the siting problem and work toward the solution in a consistent and rigorous manner.

2. The rationale and documentation necessary to justify a siting decision are provided automatically, since all aspects of the methodology — from data sources to underlying assumptions — are explicitly stated.

6.3.3. QUALITATIVE AND QUANTITATIVE

A good methodology includes *qualitative* as well as *quantitative* considerations. On the surface it appears that the siting problem is mostly quantitative. For example, a state department of environmental protection hands down quantitative criteria. These may be based on hydrogeologic parameters (such as soil permeability and distance to the water table) and geographic factors (such as distance to main roads and population centers). The criteria may be applied to potential sites by means of an objective methodology to yield a list of best sites.

Regardless of how objective the decision process attempts to be, however value judgments and other qualitative evaluations must be made. The setting of criteria implies value judgments and influences the way impacts are later evaluated. Qualitative considerations must be taken into account.

The siting methodology must transcend the objective criteria—must avoid the pitfalls of attempting to be purely objective—for three reasons. First, the nature of the siting problem requires the methodology to take into account risk and uncertainty, future impacts, and various levels of tradeoffs. These are qualitative considerations requiring subjective judgments based on the outlooks of the people involved.

Second, a good methodology must involve the public and take into account conflicting opinions. This involves tradeoffs between conflicting requirements, especially when no sites are found that meet all the important criteria. For example, there may be no sites in the State of New Jersey that are not over a major aquifer, not within a certain distance of population centers, *and* not in environmentally significant areas. If this happens, tradeoffs involving judgments as to the relative importance of each of the three criteria will have to be made. Perhaps it will be decided that a site near a city is acceptable, provided it is not over a major aquifer and the state park system is protected. Such tradeoffs will necessarily take place on a large scale in various aspects of the decision process in order to find suitable sites. The site-selection methodology must reflect the qualitative nature of such tradeoffs.

Third, siting is ultimately a political decision. It is not the purpose of the site-selection methodology to come up with hard answers but with suggested solutions. It is a tool of the decision maker in the decision process. Criteria can be used to compare the relative strengths and weaknesses of each site, but the political process makes the final decision, approving or rejecting each suggested site. To be flexible enough to be useful as a tool in the decision process, then, the site-selection methodology must include qualitative considerations.

The qualitative and quantitative aspects of a methodology are related to its formal aspects. In order to include both qualitative and quantitative considerations, a methodology must make value judgments, tradeoffs, and other decisions that cannot be put into any set objective formula. In order to be formal, the methodology must explicitly state how these decisions are made, rendering them understandable and justifiable to others.

6.3.4. INVOLVE THE PUBLIC

A good methodology must effectively *involve the public*—extensively, and early on in the siting process. The primary aim of public involvement is to find sites that are most acceptable to the public and therefore least likely to be opposed by them. This involves two things:

1. The site-selection process must be equitable—that is, fair, legal, and rational, and actually taking into account the viewpoints of affected groups. While it is impossible to make decisions that please all the affected groups, it is important to listen to them and to directly address their concerns.

2. The decisions arrived at in the site-selection process must be perceived by the public as being well made. In a sensitive area such as this, people's perceptions can be more important than reality.

The public can be involved in the decision process in many ways that promote equity and accurate perceptions. For example, affected members of the public can sit on siting commissions, affected locales can do site suitability studies, and local workshops and information distribution programs can be conducted. What should be avoided, particularly, is a paternalistic attitude by the decision makers, involving the public only in the later stages of the process or not at all. The results of such an approach are usually disastrous.

The benefits of involving the public early in the decision process are many. Not only will misconceptions be cleared up and public opposition eased, but insights will be gained to improve the quality of the final decision, and it will be much easier in the end to justify that decision. Most important, chances are improved that facilities will actually be sited.

6.3.5. WELL EXPLAINED

That a site-selection methodology and outcome be *well explained* is of central importance to the siting process. A methodology that is very strong in the other four areas can encounter serious difficulties by being weak in the fifth. While an appropriate and formal methodology that includes qualitative and quantitative considerations and effectively involves the public will most likely be *explainable,* this does not mean it will be *well explained.* Since perceptions are just as significant as reality in situations such as this, any analysis that is poorly presented is for practical purposes unsound. It will engender public opposition and conflicts that can keep even the best site from being used.

A methodology that is well explained, then, has two benefits. One is that the public will be able to understand what has occurred. This, coupled with effective public involvement, can greatly ease public opposition. The other benefit is in justification of the outcome. As previously indicated, if it cannot be clearly communicated why the best outcome is the best, it is for all practical purposes not the best. Justification is impossible unless the analysis is well explained.

This characteristic is clearly tied to the quality of the other four. An unsound analysis is unsound no matter how it is explained, but a sound analysis must be well explained in order for its benefits to be derived.

Taken together, these five are the necessary overall characteristics of a good site-selection methodology. Let us look now at the processes by which such a methodology is worked out.

6.4. PROCESS OF SITE SELECTION

The process of site selection can be defined as the stages or levels by which the number of sites considered is narrowed down from all possible ones to the few best. At each level, sites are eliminated, depending on how the criteria are applied at that level. With each subsequent level the applied criteria become more specific, the area considered becomes smaller, and the remaining possible sites become fewer.

This narrowing down is important, since it would be unacceptably costly to attempt to apply all the criteria to all possible sites, even within a given region. The levels of the process make the problem workable, eliminating unfeasible sites in the most cost-effective way.

There are many ways to define the stages or levels of a site-selection process. Generally the process will have four levels. The first three narrow down the field of candidate sites from all possible to a few best. At the fourth level the decision maker and the political process select the actual sites. The first three levels are the *regional,* the *feasible area,* and the *preferred site.* Each has its own characteristic geographic size, application of criteria, and depth of study.

The regional level defines the region of interest. In New Jersey, for example, the area of interest might be "somewhere in New Jersey" or "within 100 miles of major waste production centers." The aim is to cut out major areas that are not of interest and need not be analyzed. The method is called *preliminary screening.* The decision maker defines broadly the limits within which the sites must be found, and thereafter no other areas are considered. On this level only one or two criteria are applied and the analysis is straightforward.

Once the region of interest has been determined, smaller feasible areas must be established within which suitable sites are likely to be found. This is done by finding groups of sites that meet a broad set of criteria having two major characteristics: (1) the criteria are very important, so that the areas excluded at this level are extremely undesirable and offer little hope of producing a suitable site; (2) the criteria are easy to evaluate. The purpose is still to exclude from further analysis major portions of land as easily as possible without throwing out potential sites. An example would be excluding all areas that are over major aquifers. Since this information is readily available, it would not be difficult to cross out, on a map, the area over all major aquifers. The remaining area would be left open for consideration. Additional important and broad criteria could then be applied to further reduce this area. In New Jersey this could be done by applying some of the criteria established in section 9 of Senate Bill S-1300, which controls the establishment of hazardous waste facilities in New Jersey. Some of these criteria prohibit

> the location or operation of any new major hazardous waste facility within:
>
> (1) 500 yards of any structure which is routinely occupied by the same person or persons more than 12 hours per day, or by the same person or persons under the age of 18 for more than 2 hours per day;
>
> (2) Any flood hazard area delineated pursuant to P.L. 1969, c. 19 . . . ;
>
> (3) Any wetlands designated pursuant to P.L. 1970, c. 272 . . . ; and
>
> (4) Any area where the seasonal high water table rises to within 1 foot of the surface, unless the seasonal high water table can be lowered to more than 1 foot below the surface by permanent drainage measures approved by the department.

Criteria (2) and (3) could be used to further narrow down the number of feasible areas. These criteria are both important and (since they are delineated and designated by law) easy to determine. Criteria (1) and (4) could not be used at this level, however. Although they are important enough to be written into law, they are not easy to determine (they are site-specific and will become important at the next level). Analysis on this level is called *screening* and is carried on at a moderate depth of study.

The third level of the process is determining individual sites. It involves in-depth study, increasing as the number of sites considered is further reduced. It involves applying specific criteria to a feasible area to determine potential sites. These sites are then compared and preferences established. At this level the quality of the methodology becomes most important, since specific sites are actually being determined.

Figure 6.1 diagrams how the process is worked out. Site-selection criteria are applied more specifically as one moves from top to bottom. The more general criteria are always applied first. This is important, since it is not cost-effective to apply specific criteria to an area that could have been eliminated more simply by more general criteria. The difficulty of the analysis and the importance of the methodology also increase as one moves from top to bottom on the diagram.

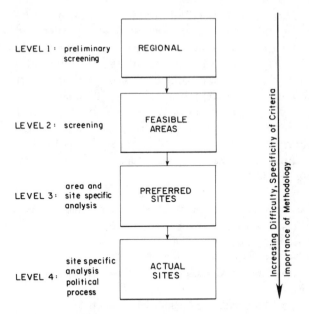

Figure 6.1.
The siting process (after Baecher, 1975).

Within each macro-level analysis a subprocess is applied that carries out the analysis in a formal manner. For different macro-levels, different aspects of the subprocess become important, but at all levels one can define five steps that represent the subprocess:

- *Step 1.* Define objectives.
- *Step 2.* Determine attributes.
- *Step 3.* Model the problem.
- *Step 4.* Review results.
- *Step 5.* Iterate.

Figure 6.2 shows the relationship between the process and the subprocess. The functioning of the subprocess is described below.

6.4.1. DEFINE OBJECTIVES

An *objective* is the aim or end result that the decision process is trying to achieve. For example, in the siting of hazardous waste disposal facilities one objective would be the protection of all major sources of water. There are different levels of objectives. Higher-level objectives, called *major objectives,* define the scope of the problem, while lower-level objectives, or

DECISION FRAMEWORK FOR SITING HAZARDOUS WASTE FACILITIES

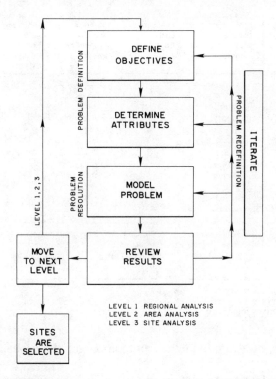

Figure 6.2.
The subprocess.

subobjectives, clarify and define the major objectives and together completely describe the problem. Major objectives are derived from the five concerns discussed earlier in this chapter. One major objective could be to "minimize adverse health and safety effects at a given site." An example of a subobjective would be to "protect the public water supply" or "minimize the contamination of water-supply wells." "Prevent all contamination of wells" and "eliminate adverse health and safety effects" would not be considered objectives, since they define a cutoff point, a way of measuring the impact. They are called *attributes.* Objectives serve only to describe the problem, while atrributes provide a way to *measure* the extent to which an objective is achieved.

The way in which major objectives, subobjectives, and attributes are arranged is called an *objectives hierarchy.* Figure 6.3 shows a partial objectives hierarchy, describing the relationship between various levels of objectives. The benefits of an objectives hierarchy are many. Most important, it helps make the problem understandable by showing specifically the way in which it is formulated. This improves clarity and consistency for those attacking the problem and aids greatly in justifying criteria and site recom-

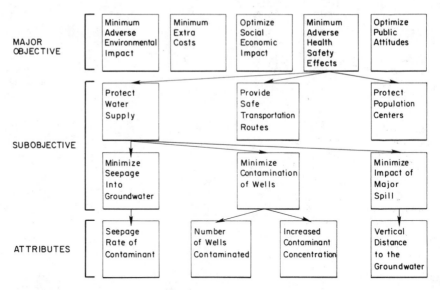

Figure 6.3.
A partial objectives hierarchy.

mendations. It also provides a way to approach the problem comprehensively.

One could start with the five concerns as major objectives and then break each down so as to clarify it with respect to the problem at hand. This process is repeated for each level until the lowest-level subobjectives become readily quantifiable. Objectives at each level should fully describe all relevant aspects of the objectives in the next higher level; if they do not, they should be reformulated. In this way the objectives hierarchy improves the formulation of the problem. It yields objectives that completely describe the relevant aspects of the problem and are at the same time nonredundant and minimal.

In order to remain in the hierarchy an objective must have three characteristics. First, it must be important in the siting process. This can be determined by asking the following questions in order (Keeney, 1980):

1. Can the objective be achieved to different degrees at at least two sites?
2. Is this difference significant relative to the other differences between the sites?
3. Is the likelihood of this difference's occurring large enough to justify its inclusion?

If the first question cannot be answered affirmatively, then the objective is achieved at all sites and is therefore not a deciding factor in site selection.

The second question becomes important for an objective such as "minimize extra costs due to road construction." If it were found, for example, that road construction costs for the most expensive site were $10,000 and for the least expensive $5,000, this difference would be considered relatively insignificant and the objective, therefore, would be considered unimportant.

Often government agencies will establish attributes criteria for siting decisions. Implicit in these criteria is an objectives hierarchy. Before analyzing the problem, it is important to "back out" of the criteria list the objectives hierarchy. The reason is that the analyst must consider not only the quantitative requirements but also the implied qualitative concerns. In addition, the hierarchy must be readily apparent, if not explicitly stated, in the public explanation of siting decisions and be open to reevaluation and reformulation as the decision process is carried out.

6.4.2. DETERMINE ATTRIBUTES

An *attribute* is a measurement of the degree to which an objective on the lowest level of the hierarchy is achieved. Collectively the attributes measure the degree to which all the objectives are achieved. Each of the lowest-level objective may have one or more attributes, depending on how specific it is. For example, a health and safety subobjective, "minimize increase in mortality due to a facility," can be measured with only one attribute "increase in mortality (in additional deaths per year)." On the other hand, a subobjective such as "minimize potential adverse impact on water supply due to major accident" may need many attributes to adequately measure it, since many types of accidents are possible and there is no direct way to measure the impact.

Attributes that directly follow from the subobjective, such as from the mortality-increase subobjective, are called *natural attributes.* Ones that do not directly follow are called *proxy* (or *surrogate*) *attributes* (Baecher, 1975). Proxy attributes are used when the primary property of the subobjective is inherently unmeasurable or the natural measure is analytically intractable.

The set of attributes should be complete, nonredundant, and operational. *Completeness* means that attributes should fully measure their respective subobjective; *nonredundancy* means that they should not double-count impacts. *Operational* means that data needed for the measurement should be appropriately obtainable. It also implies that the set of attributes is minimal, so that they can be managed in evaluating the objectives.

It is the attributes, not the objectives, that are used in the actual analysis. One must, however, understand the attributes in the context of the objectives hierarchy. Although ideally the determination of attributes is logically consistent and systematic, in practice it is inherently subjective, involving professional judgment, knowledge, and experience. This subjectiveness must be recognized on all levels of the objectives hierarchy, from application of the criteria to explaining what has occurred to the public.

6.4.3. MODELING THE PROBLEM

The first two steps of the subprocess serve to determine the questions that need to be asked and the context in which they should be answered. The *modeling* step provides the answers. For our purposes a *model* is the method by which achievement of the objectives, as measured by the attributes, is evaluated and the outcome of a given level is reached. The method may be very intuitive and casual or may be extremely complex, involving a myriad of computers and analysts.

At early stages a model might involve only an explicit statement of how a decision is to be made. As the siting process progresses and decisions become more complex, more formal and rigorous models are needed, since intuition and experience are no longer sufficient. For example, when areas have been chosen and specific sites are being compared, formal models are needed to sort out the many parameters describing each site, evaluate them by the criteria, compare the results, and determine which sites are "best" and why.

Models, however, are not intended to produce definitive answers but to be used as a tool in solving the problem. Their function is threefold. First, they serve to organize the data, to clear discrepancies in data, and to point out any important data that are missing. Second, models amplify the information obtained from a given data base, greatly increasing its amount and quality. Third, models provide a coherent and explicit method for evaluating and comparing sites, taking into account all the important inputs into the decision process and indicating which sites are best. Without such a coherent and explicit method the decision maker would have difficulty in understanding the problem at sufficient depth to find effective solutions and to defend or justify them.

Many types of models are available, each having different strengths and weaknesses. The choice depends primarily on the appropriateness of the model to the task at a given level. The time and cost involved should be in line with the results obtained. A complex, mathematical model would probably not be appropriate on the regional level, nor would a simple, more intuitive model in comparing specific sites.

6.4.4. REVIEW RESULTS AND ITERATION

The first three steps of the subprocess at each level yield a preliminary decision that narrows down the size of the area and the number of sites to be considered. A review of all that has occurred in the first three steps is necessary in order to assure that the objectives are defined, the attributes determined, and the problem modeled and evaluated in the best possible way so as to produce the best possible decisions. A full review can reveal omissions, errors in computation or judgment, and other inadequacies. Important objectives may have been left out and unimportant ones left in;

attributes may be inadequate or poorly measured, models inappropriate, and the entire process too informal or too qualitative.

A second reason for review is that the problem is constantly being redefined. Since learning is a continuous process, it is impossible to completely understand a problem at the outset. The objectives hierarchy may not be fully understood until after the problem has been modeled a number of different ways. In addition, new information is gained, values change, and perceptions of the problem mature as the decision process progresses. Review is needed to take these factors into account. For example, after the problem has been modeled, objectives that were thought to be important may be seen to be relatively unimportant by the criteria previously established. Differences in the degree to which they are achieved may be small at all sites, and the objective must therefore be excluded.

Review is important, too, so that the decisions made can be properly justified. The process of formally questioning and checking for errors to make sure that the decisions are relevant to the present understanding of the problem helps make explicit why decisions were made the way they were. In addition, explaining what has occurred leads to an understanding of the limits of the problem, clarifying the conditions under which the decision is relevant. In operations research this is often called *sensitivity analysis*. Sensitivity analysis should always be done, since it makes clear the context in which the decisions apply. If this context changes — such as with different assumptions or professional judgments — sensitivity analysis shows whether or not the solutions change.

Finally, review is necessary for the sake of public participation. While participation is needed at all stages of the process, public review of what has occurred at each level is particularly important — for two reasons. One, it helps defuse public opposition to a final decision, since the public was given the opportunity to be aware of preliminary decisions as they were made. The fact that preliminary decisions are disclosed and available for review helps the public perceive the final decision as being well thought out and not just arbitrarily imposed. Two, if public opinion is in fact taken into account, it gives the public an input into the decision process and provides an opportunity for changing what has been done to avoid unnecessary public opposition.

While review of the decision process should be constantly taking place, it is particularly valuable to have a point at which the entire process up to that point can be reviewed by all relevant parties. Questions that need to be answered include:

1. Is the problem redefined? (If so, where?)
2. Was the problem formulated properly?
3. Are the results reasonable? Appropriate? Justifiable?

4. Are the five concerns adequately dealt with?
5. Does the process have the characteristics of a good methodology?

Once these questions have been considered, one more needs to be answered: "Is iteration necessary?" In other words, "Is it necessary to change the way the problem has been defined, formulated, modeled, or evaluated and to find a new solution so that the above questions can be answered in the affirmative?" If the answer is "yes," then the process is redone with necessary changes and the question is asked again. Only when this question is answered "No" are we ready to move to the next level, where the subprocess begins over again.

6.5. MAKING A FINAL DECISION

The end result of the process and subprocess is a list of recommended sites and a justification of why they are recommended. The justification should describe how the chosen sites are best in terms of the major environmental, economic, socioeconomic, health and safety, and public attitude objectives; it should also compare their relative strengths and weaknesses in these five areas. In addition, the justification should show why these sites were chosen and all others eliminated. This means that it must describe the decision process, indicating how and why each preliminary decision was made. It is then the job of the decision maker to choose the actual sites on which facilities are to be built.

CHAPTER SIX REFERENCES

BAECHER, G. B., *Balancing Apples and Oranges: Methodologies for Facility Siting Decisions,* International Institute for Applied Systems Analysis Research Report (Schloss Lasenburg, Austria, September 1975), p. 6.

KEENEY, R. L., *Siting of Energy Facilities,* Academic Press, New York, 1980, 413 pp.

NEW JERSEY SENATE, Bill S-1300, Sections 3k and 9a, pp. 5 and 14.

UNITED STATES ENVIRONMENTAL PROTECTION AGENCY, "Everybody's Problem: Hazardous Waste," SW-826, USEPA Office of Water and Waste Management, Washington, DC, 1980, p. 13.

CHAPTER SEVEN

APPLICATION OF DECISION ANALYSIS FOR SITING HAZARDOUS WASTE FACILITIES

7.1. INTRODUCTION

The site-selection process for hazardous waste, treatment, storage, and disposal facilities requires judgments based on the values and priorities of the decision makers and important parties-at-interest. Many site-selection strategies attempt to avoid addressing these judgments by remaining "objective." Such objective methodologies, however, deny the very nature of the siting problem and serve only to obscure the judgments that inherently must be made. Other strategies address judgments by defining them apart from the actual decision process. This is the case with many site ranking methods. Values are "defined" by "assigning" weights (a measure of the relative importance of siting objectives used to model judgments) based on technical expertise and "expert" judgments. These weights are imposed on, rather than derived from, the decision process. This leads to siting decisions that may not be appropriate to the given area or justifiable by the decision makers (Bass, 1982).

Site-selection methods are therefore needed that explicitly incorporate the actual values inherent in the siting process and lead to decisions that are both appropriate and justifiable (Bass, 1982). One site-selection approach that attempts to do this is *decision analysis.*

This chapter shows how decision analysis can be used to identify the most suitable land emplacement facility site for hazardous waste based on the site's hydrogeologic parameters. Decision theory provides a logical and systematic framework to structure objectives and to evaluate and rank alternative potential sites. The proposed framework aggregates the impacts of geology and hydrology into a single indicator that measures a site's desirability. The derived utility function, which indicates preferences toward site characteristics, is applied to Sussex County, New Jersey. This location was chosen because of its complex geology and comprehensive data base. If the process of identifying, selecting, and ranking potential sites in such a heterogenous area can be performed, application of the same process to a more homogenous area should be an easier task.

A formal site-selection methodology, such as decision analysis, has three components within which the general siting methodology developed earlier fits. The case study is developed around these components, which are as follows:

1. *Defining the objectives.* Objectives are the major goals and concerns of the site-selection process.

2. *Determining attributes.* Attributes measure how well the objectives are being achieved. Attribute levels for candidate sites are based on models and technical judgments.

3. *Determining weights.* The weights measure the relative importance of the objectives (as measured by their attributes) and are based on subjective judgments of decision makers, environmental commissions, and so forth.

7.2. SUSSEX COUNTY SITING STUDY

The specific siting problem involves Sussex County, New Jersey. This study considered only the hydrogeological parameters and as such is incomplete. A complete study would include economic, social, and political objectives and measures, as discussed in Chapter 6.

Furthermore, the consequences at each candidate site were evaluated by means of very approximate models and parameters. The purpose of the study is to illustrate the approach and not to serve as a final decision tool. Nevertheless, the results are useful and corroborate the findings by Sussex County's consultants in a solid waste siting study.

7.2.1. NEW JERSEY GEOLOGY: SUSSEX COUNTY

Two major physiographic regions—the Atlantic Plain (Continental Shelf and Coastal Plains) and the Appalachian Highlands (Piedmont, New England, and Valley and Ridge Provinces)—make up New Jersey. The relief map of New Jersey is shown in Figure 7.1.

Southeastern New Jersey has a general elevation less than 500 feet and

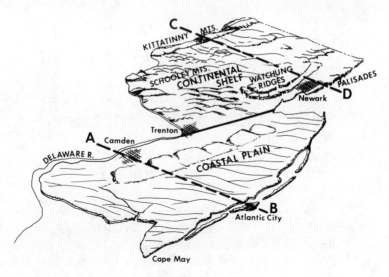

Figure 7.1.
Relief map of New Jersey (after Hunt, 1974).

lies in the Atlantic Plain. It consists of Cretaceous shales (Ksh), Cretaceous sandstones (Ks), and Tertiary (T) and Quarternary (Q) formations of mostly marine origin, intertongued with continental deposits, which slope slightly northeast. Ridges parallel to the coast are formed by the more resistant formations such as Cretaceous sandstone (Ks). Note that Atlantic City lies on a barrier beach, as shown in cross section *AB* in Figure 7.2.

The geology of northeastern New Jersey is more complicated than that of the southeast. The Triassic (Tr) belt of block mountains between Trenton and Newark Bay divides the coastal plains from the Appalachian Highlands. The elevation of the northwestern part of the state varies between 500 and 2,000 feet, owing to rolling uplands of the Appalachian Highlands. Earlier geologic activity has folded the Lower Paleozoic [Cambrian (E), Ordovician (O), and Silurian (S)] formation, which is mostly of marine origin. See cross section *CD* in Figure 7.3.

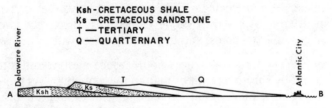

Figure 7.2.
Cross section *AB*, Coastal Plains (after Hunt, 1974).

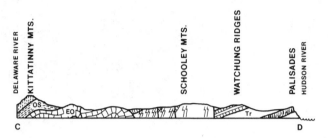

Figure 7.3.
Cross section *CD*, Continental Shelf (after Hunt, 1974).

Sussex County, the northernmost county in New Jersey, has a land area of 526 square miles. It is bounded on the northeast by New York State, on the northwest by the Delaware River and Pennsylvania, on the southwest by Warren County, and on the southeast by Passaic and Morris Counties.

7.2.2. IDENTIFYING CANDIDATE SITES

The study area of Sussex County is comprised of 526 square miles. Not all locations in the county are candidate sites. For a variety of reasons, large areas of the county can be excluded from the siting process. The process of reducing the potential sites is described in Chapter 6 (Section 6.4). Excluded areas included the following:

Major Aquifer and Critical Recharge Zones: These areas comprise potentially high-yielding carbonate rocks (Oontelaunee, Epler, Rickenbach, Undifferentiated Kittatinny, Allentown, and Leithsville formations) and aquifers defined by formations that consist mostly of sands and gravels. Sussex County officials considered the value of these major water sources to be so high that all candidate sites should be excluded.

Size: Size of land emplacement facility is a physical and an economic constraint. A site must be large enough to accommodate the volume of incoming hazardous waste for a period in excess of twenty years and to include buffer zones and access roads. For example, a minimum of 200 acres is required for a twenty-year planning period. This period depends on the degree and efficiency of resource recovery. One can consider that the smallest area that may be developed cost-effectively must be at least .1 square mile (64 acres) and at least 1,000 feet wide.

Slope: The slope influences both surface and groundwater flow, because water flows from a point of high to low potential. As groundsurface slope increases, a more rapid decrease in elevation head occurs, causing an increase in velocity and water discharge. A facility at a high relative elevation, such as the top of a hill, will receive water only from direct rainfall, but leachate from the site may travel long distances. A facility in an area with little or no slope at the base of a hill will collect its direct rainfall plus runoff from the surrounding region. Since this area is usually at a low relative potential, leachate will probably not migrate far. A facility that is on neither a local high nor a local low will have to deal with problems of surface and subsurface runoff from an area upslope and of controlling potentially contaminating runoff from the facility.

For these reasons, sites having slopes greater than 22 percent were excluded from consideration. It was felt that a facility could be securely constructed with little difficulty if the slope of the terrain was less than 22 percent.

A significant portion of the county remained after the excluded areas were removed. As a first cut, and as a comparison to a recently completed siting study described in Section 7.3, ten candidate sites were selected from the remaining area for further analysis.

7.2.3. SPECIFYING THE OBJECTIVES AND ATTRIBUTES

Any siting procedure must specify its objectives. In the Sussex County study, the greatest concern was the impact to current or future water supplies from leachate contamination.

While "secure" landfills are claimed to exist, current investigations have shown that many (potentially all) may allow off-site migration of contaminants. Over long time spans (100 to 500 years), significant leachate movement is probable.

In this study the objective was to minimize the movement of the contamination plume off-site, given that leakage may occur. Fulfilling this objective will also fulfill the objective to minimize water-supply contamination and other impact objectives. Attributes measure how well an objective is being met. Many attributes can be used to measure a particular objective.

Derivation of the final hydrogeological attributes used to measure the objective to minimize contaminant migration is an iterative process that involves many steps. It is necessary to understand the characteristics that influence pollutant migration and travel time so as to determine which attributes are more important. Appendix A discusses in detail the hydrogeologic parameters important for siting land emplacement facilities.

Six hydrogeologic attributes were used to structure the decision-analysis problem of Sussex County. Their ranges were determined for Sussex County and were used to evaluate the degree to which alternative sites meet the objective. Table 7.1 gives the attributes and their ranges and units of measurement. The range is that found over Sussex County.

Table 7.1 Summary of Attributes and Ranges

Attribute	Range Worst	Range Best	Unit of Measurement[c]
x_1: Shortest travel time to major aquifer or critical recharge zone	0	140	years
x_2: Shortest travel time to existing well	0	80	years
x_3: Shortest travel time to surface water body	0	80	years
x_4: Shortest distance to fault, dike, natural cavernous areas, or active/inactive mines[a]	0	13	miles
x_5: Size[b]			Exclusionary
x_6: Slope			Exclusionary

[a] Kittatinny limestone yields large quantities of water and tends to have solution channels.

[b] Size is a physical and an economic constraint. The question of who uses the land emplacement facility plays a crucial role in determining its size and life expectancy. See paragraph on "Size" in Section 7.2.2.

[c] Units of measurement for attributes x_1, x_2, x_3, x_4, and x_5 have objective rather than subjective indices. There are difficulties associated with defining subjective attribute scales. A subjective index must be developed for an attribute for which no objective scale exists. It may be done by establishing a scale from 0 to 100, where 0 represents the worst and 100 represents the best level. For example, surface soils could be categorized with the associated numerical values as 0 for no potential (sands and gravels less than 1 foot), 20 as bad (silty soil layers mixed with sand 1 to 5 feet in thickness), 40 as fair (irregular lenses of clay, silt, and sand), 80 as very good (mixed layers of silts and clays 5 to 10 feet in thickness), 100 as excellent (continuous, very low-permeability clay, such as montmorillonite, greater than 10 feet in thickness).

7.2.4. EVALUATING SITE IMPACTS

Given a set of impacts for each alternative site, the next step is to evaluate the impacts' desirability. For notational ease, let us define $X = (x_1, x_2, x_3, x_4)$ as the set of impacts at a site, where x_1 is the specific value of attribute 1 and so forth. For example, for one of the alternative sites (A-43) the impact set $X = (23.5, 80, 80, 0.678)$. Since attributes x_5 and x_6 are exclusionary (that is, they are used to eliminate a site from further consideration), they do not appear.

Multiattribute utility theory provides the methods and procedures to determine whether a set of impacts is more or less desirable than another set. A utility function U is assessed that assigns a value $U(X)$ to the impact set X. The utility function has two important properties:

1. If $U(X')$ is larger than $U(X'')$, then the impact set X' is preferred to X''. Here X' and X'' are sets of impacts at alternative sites. Thus site X' is preferred to site X''.

2. When uncertainty exists in X, the expected value of U is the appropriate index upon which alternative sites are ranked.

The reader is referred to Raiffa (1968) and Keeney and Raiffa (1976) for further discussion of the details. The utility function used to evaluate sites incorporates the preferences of local environmental planners and citizens interested in the siting decision. Evaluating utility functions requires skill.

When the utility function represents the preferences of a group, additional difficulties arise. The reader can appreciate that, as a first step, the utility function serves as a focus point for public discussion for siting and for identifying sites that appear to be significantly superior.

There are essentially five steps in the evaluation of a utility function:

1. Introducing the terminology and ideas.
2. Determining the general preference structure.
3. Assessing single-attribute utility functions.
4. Evaluating scaling constants.
5. Checking for consistency and reiterating.

The theory and method of assessing a multiattribute function are explained in greater depth in Keeney and Raiffa (1976), Keeney (1977), and Keeney and Wood (1977).

Introducing the Terminology and Ideas: The dynamic interaction process of assessing a utility function is based on a series of questions by the assessor about the decision maker's preferences (Keeney, 1977). Numerous meetings were held with Lyn Dabagian to discuss the attributes and general preference structure. As the Senior Environmental Planner for the Sussex County Department of Planning, Conservation, and Economic Development, Ms. Dabagian coordinates the "208" Water Management Plan for the county. So as a first step we familiarized her with the applicability of multiattribute utility theory to the siting problem.

Determining the Qualitative Preference Structure: The second step is to verify the preferential and utility independence assumptions to determine the function's qualitative structure, which would later be quantified. If these assumptions prove to be reasonable, then the overall utility function, $U(X)$, can be described as a simple functional form of four one-attribute utility functions.

That is, the mathematical form of the overall utility function $U(X)$ is either additive:

$$U(X) = \sum_{i=1}^{4} k_i u_i(x_i) \qquad (7.1)$$

or multiplicative:

$$1 + KU(X) = \prod_{i=1}^{4} 1 + Kk_i u_i(x_i) \qquad (7.2)$$

where $u_i(x_i)$ is a one-attribute utility function for attribute i. The scaling weights k_i have the range $0 \le k_i \le 1$, and the constant K has the range $K > -1$. All the utility functions, $U(X)$ and $u_i(x_i)$, are scaled from 0 to 1.

In-depth analysis of Ms. Dabagian's preference resulted in assessing her utility function for the four attributes. These are shown in Figure 7.4. Furthermore, by comparing special sets of site characteristics, it was determined that a multiplicative form for her overall utility function was most appropriate.

Evaluating Scaling Weights: The first step in determining the scaling factors $k_i = k_1, k_2, k_3, k_4$ is to ordinally order the scaling constants. The easiest approach is to ask: "Assume all the attributes are at their worst levels. Now if only one attribute could be raised to its best level, which one would be preferred?" (Keeney and Wood, 1977). The attribute chosen has the highest scaling-factor value.

Following this approach, the following order was determined:

$$k_1 > k_2 > k_3 > k_4$$

Assessing tradeoffs between pairs of attributes allowed us to determine sets of equations that give the relative scaling factors. The five weights were solved, simultaneously, from five tradeoff equations to yield

$$k_1 = .8066, \quad k_3 = .4033$$
$$k_2 = .4606, \quad k_4 = .2307$$
$$K = -.9298$$

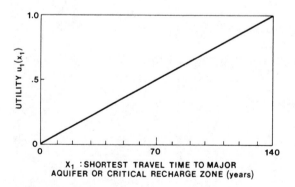

Figure 7.4a.
Utility curve for shortest travel time to major aquifer or critical recharge zone.

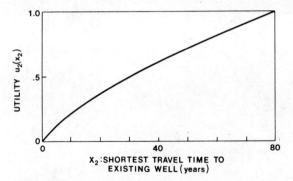

Figure 7.4b.
Utility curve for shortest travel time to existing well.

These values are substituted into the standard multiplicative multiattribute utility function given in equation (7.2). The first four weights give the relative importance of the first four attributes of Table 7.1.

7.2.5. EVALUATING ALTERNATIVE SITES

Application of the multiattribute utility function to the test area, Sussex County, was possible only through an interactive computer graphics program. The program assesses the utility of any location within the county and excludes areas with a slope greater than 22 percent. The four attributes were:

1. Shortest travel time to major aquifer or critical recharge zone.
2. Shortest travel time to existing well.

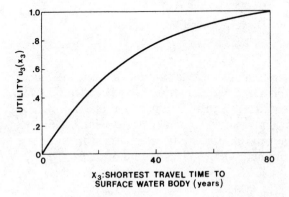

Figure 7.4c.
Utility curve for shortest travel time to surface water body.

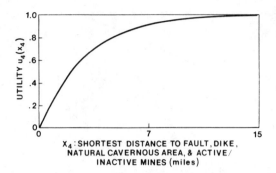

Figure 7.4d.
Utility curve for shortest distance to fault, dike, natural cavernous area, and active/inactive mines.

3. Shortest travel time to surface water body.

4. Shortest distance to fault, dike, natural cavernous area, and active/inactive mines.

These attributes were calculated from basic topographic and geologic data, obtained from a master data base. The data base included seven digitized maps, which, except for the water-table contour map, are presented in Figures 7.5 through 7.10:

1. Average Hydraulic Conductivity Map.
2. Water-Table Contour Map.
3. Aquifer Contamination Susceptibility Map.
4. 1974 Well Location Map.
5. Surface Water Bodies Map.
6. Fault, Dike, Cave, and Mine Map.
7. Slope Exclusion Map.

Numerous maps were accumulated from the Sussex County Department of Planning, Conservation, and Economic Development, the New Jersey Department of Environmental Protection (Geological Survey) and Princeton University's Geology Department. These maps played an important part in this case study. Sometimes an entire map was used, other times a combination or interpretation of maps.

7.2.6. COMPARING ALTERNATIVE SITES

Ten alternative sites were initially chosen for comparison. These sites were identified by Converse/Tenech Consultants in a sanitary landfill siting study

APPLICATION OF DECISION ANALYSIS FOR SITING HAZARDOUS WASTE FACILITIES 131

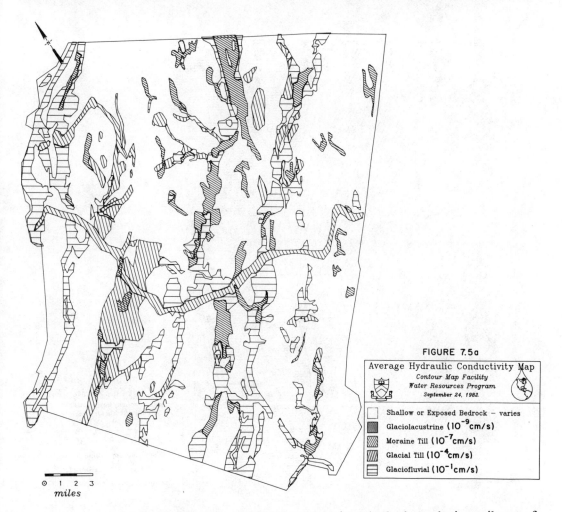

FIGURE 7.5a

Average Hydraulic Conductivity Map
Contour Map Facility
Water Resources Program
September 24, 1982.

Shallow or Exposed Bedrock – varies
Glaciolacustrine (10^{-9} cm/s)
Moraine Till (10^{-7} cm/s)
Glacial Till (10^{-4} cm/s)
Glaciofluvial (10^{-1} cm/s)

for Sussex County. Table 7.2 summarizes the hydrogeologic attributes of the ten sites; their locations are given in Figure 7.11.

The utility values for the ten sites are given in Table 7.3. It should be stressed that these results are based on a fairly rough analysis and simplified assumptions concerning the flow velocities. Field investigation can now be carried out on the top-ranked sites to verify their attribute levels.

Sites A-43, A-44, and A-47 are ranked by Converse/Tenech as the top three out of the ten sites based solely on hydrogeologic criteria (see Section 7.3). Application of the multiattribute function to the same ten sites yields the same results, although the attributes considered and the method are different. The sites with lower utility and given lower total points by Converse/Tenech vary more. For example, using the utility function, site A-11 ranked ninth, but with Converse/Tenech it ranked fourth.

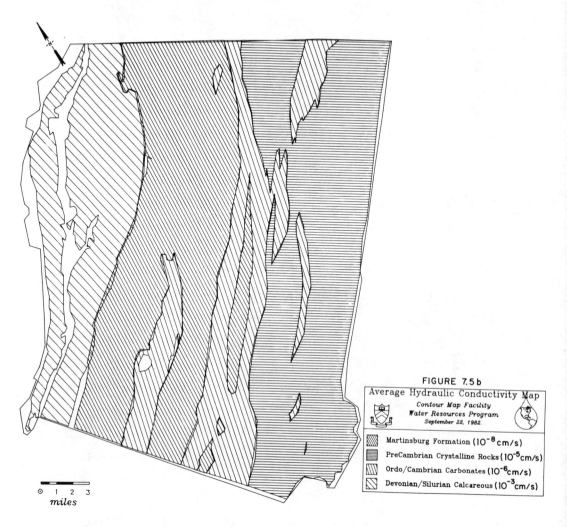

FIGURE 7.5b
Average Hydraulic Conductivity Map
Contour Map Facility
Water Resources Program
September 22, 1982.

Martinsburg Formation (10^{-8} cm/s)
PreCambrian Crystalline Rocks (10^{-5} cm/s)
Ordo/Cambrian Carbonates (10^{-6} cm/s)
Devonian/Silurian Calcareous (10^{-3} cm/s)

7.2.7. CASE-STUDY SUMMARY

Table 7.4 lists the advantages and disadvantages of applying decision analysis to the hazardous waste siting problem.

Many analysts may find decision analysis difficult to apply. The correct structuring of the problem and the assessment of tradeoffs among attributes require skill and experience. On the other hand, the approach is a formal, rigorous one that fulfills the siting methodology requirements laid out in Chapter 6. Decision analysis has also proven effective in actual siting cases, providing important insights not available by other techniques.

APPLICATION OF DECISION ANALYSIS FOR SITING HAZARDOUS WASTE FACILITIES 133

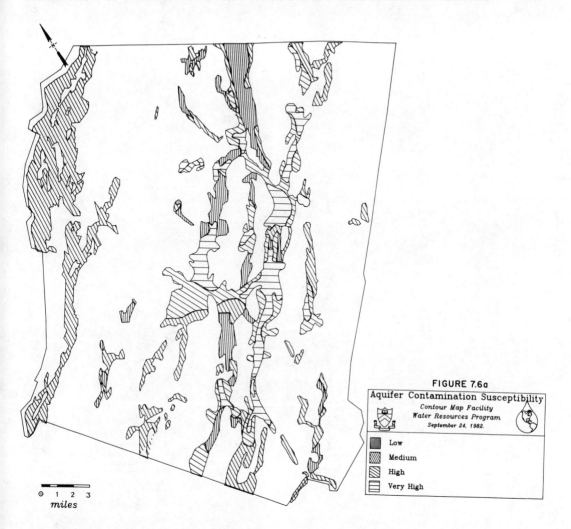

FIGURE 7.6a
Aquifer Contamination Susceptibility
Contour Map Facility
Water Resources Program
September 24, 1982.

Low
Medium
High
Very High

7.3. CONVERSE/TENECH SITING STUDY FOR SUSSEX COUNTY

Converse/Tenech (1981) developed for Sussex County a four-phase process to identify and select potential sanitary landfill sites to meet the Sussex County Solid Waste Management Plan. The goal was to find one site that would undergo detailed hydrogeologic investigation before final approval and construction.

The incorporated methodology is extremely interesting when viewed within the framework of Chapter 6. Public participation was used exten-

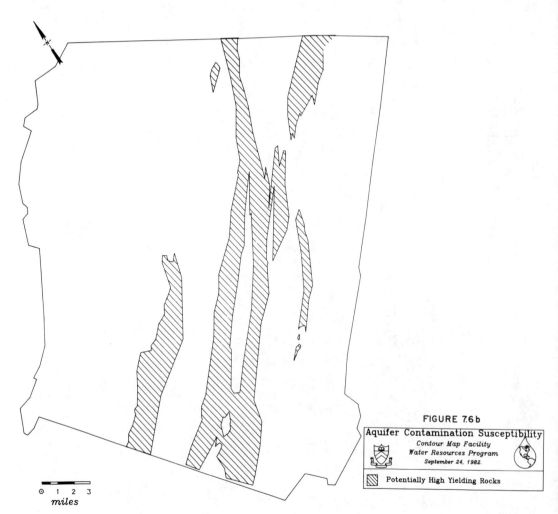

FIGURE 7.6b
Aquifer Contamination Susceptibility

sively in both defining and evaluating the siting parameters. As discussed later, essentially only hydrogeological parameters were used in the ranking study. On the other hand, the final decisions ultimately were influenced more by social and political considerations which were not part of the formal procedure — a clear inconsistency. If in the end these considerations were important, then they should be included explicitly and the sites reevaluated. The public, through four major groups, participated with Converse/Tenech in developing the study. The four groups were the Solid Waste Advisory Council (SWAC), an oversight group, local municipalities, and the Sussex County Board of Chosen Freeholders. SWAC, representing Sussex County's 24 municipalities, set up the methodology and guidelines for the study and created the oversight group. The oversight group selected

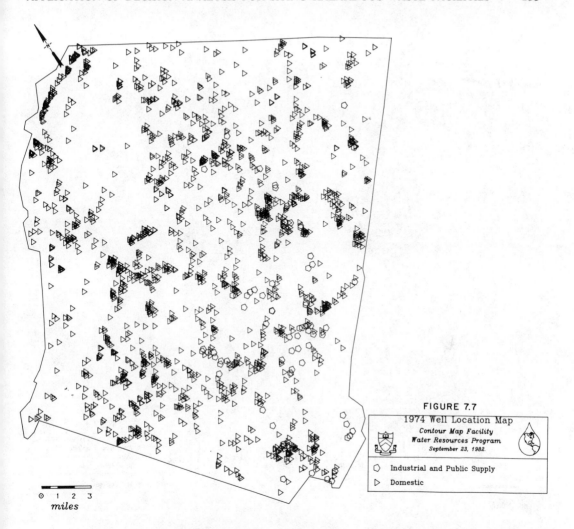

FIGURE 7.7
1974 Well Location Map
Contour Map Facility
Water Resources Program
September 23, 1982.

○ Industrial and Public Supply
▷ Domestic

the siting parameters and their appropriate weights that were used to evaluate and rank the candidate sites.

The local municipalities provided information and data concerning potential sites to Converse/Tenech, which coordinated and dealt with the technical aspects. The Board of Chosen Freeholders was to make the final decision on which sites would be taken to Stage II, where detailed on-site analysis would be carried out.

In Stage I, the goal was to reduce the county down to first ten, then four, then one site. Four major steps were involved:

- *Step 1.* Initial screening of Sussex County by applying broad exclusionary criteria.

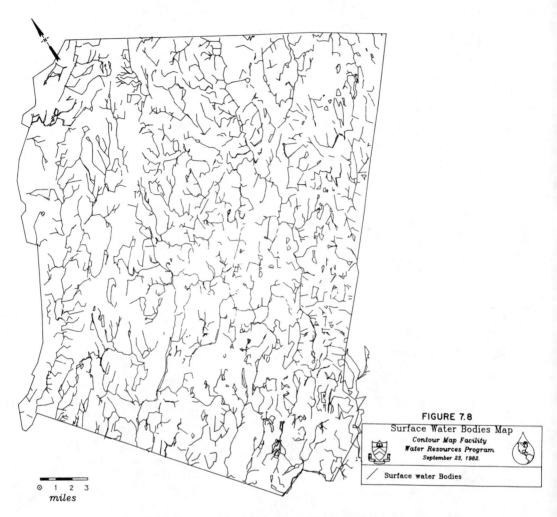

FIGURE 7.8
Surface Water Bodies Map
Contour Map Facility
Water Resources Program
September 23, 1982.

- *Step 2.* Further screening by applying hydrogeotechnical exclusionary criteria and by ranking the remaining sites based upon hydrogeological criteria. Ten sites emerge from step 2.
- *Step 3.* Further ranking by applying an additional fourteen hydrogeological and environmental criteria. Four sites emerge from step 3.
- *Step 4.* Further ranking based upon field investigations such as borings and soil sampling. The most preferred site is identified for detailed analysis under Stage II.

The broad exclusionary criteria of the first step eliminated from further analysis about 50 percent of the county—such areas as wetlands, flood-

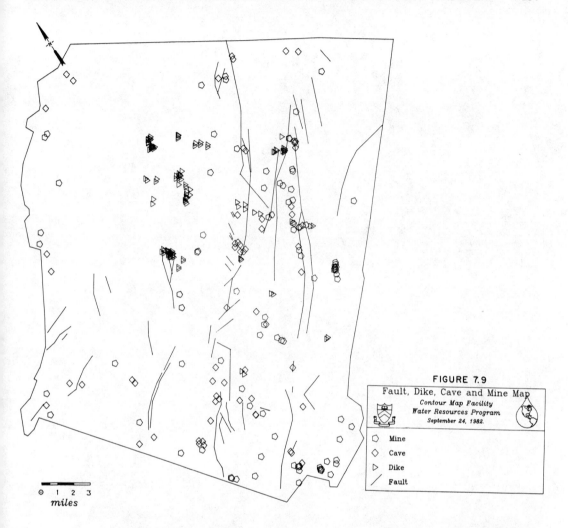

FIGURE 7.9 Fault, Dike, Cave and Mine Map

plains, archaeological and historical sites, dedicated lands, habitats for threatened or endangered species, and developed land.

Step 2, in which hydrogeological criteria were applied, excluded another 40 percent of the county—areas of seasonal high water table, areas prone to subsidence, areas of earthquake activity, sites with greater than 22 percent slope, and sites less than 64 acres or less than 1,000 feet wide. Applying the exclusionary criteria left ninety-one potential sites.

SWAC identified six major hydrogeological criteria it felt were important. These criteria along with their weights are as follows: soils (74), geology (74), groundwater (99), monitoring aspects (28), soil-cover material (34), and slope (41).

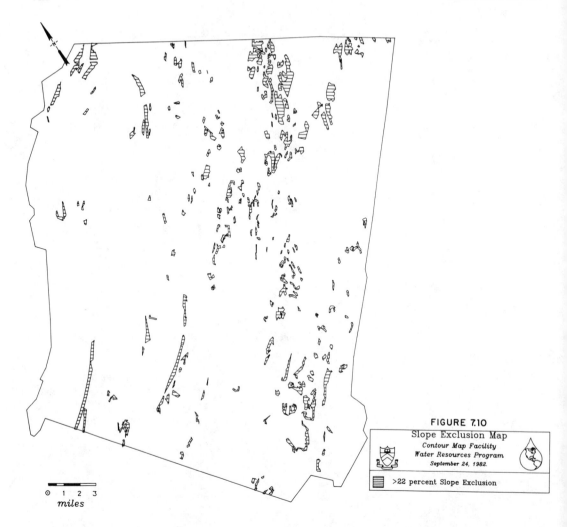

FIGURE 7.10
Slope Exclusion Map
Contour Map Facility
Water Resources Program
September 24, 1982.

>22 percent Slope Exclusion

The soil, geology, and groundwater criteria were broken down into subcriteria. These subcriteria, along with their weights, are presented with the major criteria in Figure 7.12. Notice that the creation of subcriteria and the adding of the impacts has made permeability, aquifer yield, and aquifer use less important than soil-cover material and site slope—a questionable situation.

To evaluate sites, the weights are multiplied by a rating of 1, 2 or 3, depending upon whether the criteria are found to be good, moderate, or fair at a particular site. Any criterion found to be poor eliminates that site from further consideration. The result of step 2 is the identification of the top ten sites, which are then used in step 3. The rankings of the top ten sites are

APPLICATION OF DECISION ANALYSIS FOR SITING HAZARDOUS WASTE FACILITIES   139

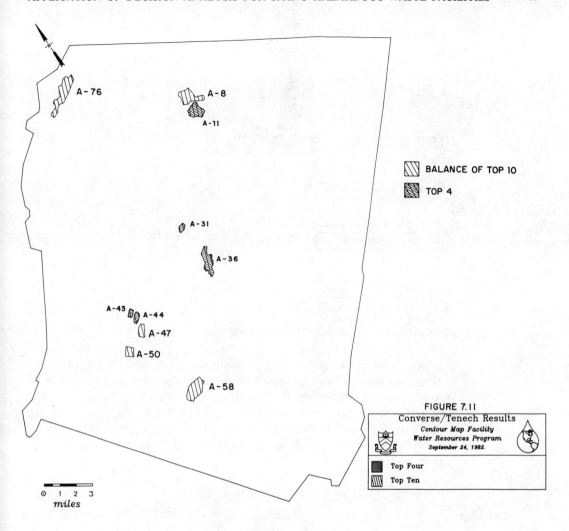

FIGURE 7.11
Converse/Tenech Results
Contour Map Facility
Water Resources Program
September 24, 1982.

given in Table 7.3, presented earlier; their locations are shown in Figure 7.11.

Step 3 established an additional fourteen criteria to reduce the ten sites down to four. These criteria, with their weights, are: land use (57), site access (43), utilities (45), soils (83), surface and groundwater hydrology (94), geology (85), topography (70), flora and fauna (42), air quality (59), noise (54), odors (65), aesthetic impact (54), sensitive areas (59), and hazards (40). The top four sites, based upon the composite score for the ten candidate sites, were further analyzed in step 4. Under step 4 field investigations were carried out to determine the most suitable site.

After the top ten sites were located within the county based on hydro-

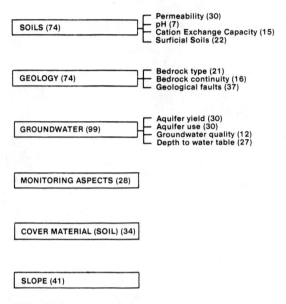

Figure 7.12.
Step 2 criteria and weights for Converse/Tenech study.

geologic criteria, site A-36 ranked seventh out of ten, while site A-44 ranked in the top four. Therefore, based on the described process, only site A-44 would proceed to the Stage II of the analysis, which incorporates nonhydrogeologic parameters. However, the Board of Chosen Freeholders decided that sites A-43 and A-44 should be considered as one, since they are very close to one another, and that site A-36 should move from rank seven to four.

Table 7.2 Attribute Levels for the Ten Alternative Sites

Site Number	Attribute Level[a]			
	x_1	x_2	x_3	x_4
A-44	89.31	80	80	.66
A-31	140.00	.48	3.40	.03
A-43	23.46	80	80	.68
A-47	.07	80	80	.35
A-58	.06	.07	80	.06
A-50	.05	1.43	11.25	.06
A-76	.10	.13	7.9	.71
A-8	.21	.13	1.88	.56
A-11	.09	.03	.08	.47
A-36	.03	.05	.08	.07

[a] x_1 = shortest travel time to major aquifer or recharge zone (years)
  x_2 = shortest travel time to an existing well (years)
  x_3 = shortest travel time to surface water body (years)
  x_4 = shortest distance to a fault, dike, natural cavernous area, or active/inactive mine (miles)

Table 7.3 Utility Values and Converse/Tenech Rankings for the Ten Alternative Sites

Site Number	Utility Value	Converse/Tenech Rank
A-44	.834	3
A-31	.773	4
A-43	.712	1
A-47	.662	1
A-58	.384	9
A-50	.136	5
A-76	.134	5
A-8	.030	10
A-11	.004	4
A-36	.002	7

Sussex County had not reached a final decision as of 1982. Site A-36 is considered the top choice, even after field analysis detected major geologic problems. For example, an unmapped fault zone and joint planes ran across the site, which is a possible recharge zone to a major adjacent aquifer. However, social impact and political reasons such as industrial zoning, low population density, location within the middle of the county, and location at a cross section between two major highways outweighed the hydrogeologic problems. Site A-44 has favorable hydrogeologic characteristics, and the combined site A-43/44 scored the highest among the 91 areas that emerged from step 1.

7.4. THE CONJOINT APPROACH TO SITING*

Conjoint analysis utilizes a simple yet powerful statistical method to measure and analyze the judgments involved in deciding among a number of less-than-ideal alternatives. These alternatives are in the form of hypothetical sites described on a card in a deck of cards describing various hypothetical sites. Some sample cards are shown in Table 7.5. Each card represents a single site, described in terms of the major objectives of the siting process

Table 7.4 Advantages and Disadvantages of Decision Analysis

Advantages	Disadvantages
Measures values in the decision process	Very complex, requiring considerable expertise to administer as well as considerable effort from the decision maker
Very rigorous	
Excellent when there is a single decision maker	
Proven effective in actual siting cases	Only one person's values are explicitly considered in each analysis
	Not applicable for multiple decision makers, such as Siting Board

* From Bass (1982).

Table 7.5 Hypothetical Site Cards

Description of Site #1

Impact on environmentally significant areas: Significant adverse impact over 50 percent of area.

Facility location: No sensitive land uses within one-half mile of the facility.

Chance of transportation accident: Higher than average.

Impact on local economy: Adversely impacts (e.g., property values down, local business hurt).

Transportation costs to facility: Very high.

Chance of significant delay due to local opposition: Moderate.

Description of Site #2

Impact on environmentally significant areas: Significant adverse impact over entire area.

Facility location: No sensitive land uses within one-half mile of facility.

Chance of transportation accident: Higher than average.

Impact on local economy: Little or no impact.

Transportation costs to facility: Moderate.

Chance of significant delay due to local opposition: Low.

Description of Site #3

Impact on environmentally significant areas: No area affected.

Facility location: One or more sensitive land uses (e.g., school, hospital) within one-half mile of the facility.

Chance of transportation accident: Average.

Impact on local economy: Little or no impact.

Transportation costs to facility: Very high.

Chance of significant delay due to local opposition: Low

(e.g., protect environmentally sensitive area, moderate impact, no impact). Hypothetical sites differ, therefore, in the degree to which each of the objectives is achieved. In addition, the site descriptions are carefully formulated (based on a computer-generated algorithm) so that the most information possible can be gained from subsequent analyses.

Once the cards have been formulated and are well understood by the decision makers, the decision makers are asked to rank a deck of hypothetical sites from most to least acceptable (highest to lowest overall utility). Since none of the choices are "ideal," this requires the decision maker to make difficult tradeoffs among objectives. The ranked set of cards are then analyzed (with computer assistance), producing as primary output:

 1. A measurement of the relative importance of each objective (weights) for each decision maker; and

 2. A measurement of the relative importance of levels within each objective (utility curves for that objective).

Table 7.6 Advantages and Disadvantages of Conjoint Analysis

Advantages	Disadvantages
Measures values in the decision process	Less rigorous than decision analysis
Relatively simple, both to formulate and analyze, and requires reasonable effort from the decision makers	Unproven in the area of site selection, although proven in similar areas
Values of many people can be considered	
Applicable to single and multiple decision makers	

This approach can be used to rank sites, build consensus (by identifying subgroups and individual differences), characterize important parties-at-interest, and defend siting decisions on the basis of a more objective understanding of the values and tradeoffs involved.

7.4.1. COMPARISON TO DECISION ANALYSIS

Conjoint and decision analysis both measure the values inherent in the decision process. The importance and benefits of doing so need not be discussed here. The advantages and disadvantages of decision analysis were given in Table 7.4 and of conjoint analysis in Table 7.6. In general, decision analysis is more rigorous than conjoint analysis and is ideally suited to a single decision maker. However, decision analysis is an involved process, may lead to larger errors (Keeney, 1980) and requires a great deal of time and expertise to use.

Conjoint analysis, on the other hand, is less rigorous and is geared toward multiple decision makers (a situation more likely to occur in the siting of hazardous waste disposal facilities). In addition, it is relatively simple to use, requiring much less effort and explicit understanding on the part of the decision makers. If both methods are properly used, decision analysis will probably produce more precise results, but conjoint analysis will probably produce results nearly as good with much less effort.

7.5. CLOSING REMARKS

Chapter 6 laid forth a framework for making siting decisions that emphasized formalism. It is argued that such a decision framework is needed to bring decision makers together with analysts and with the public. Any other framework may lead to the exclusion of good sites from consideration or public rejection of chosen sites. Decision analysis is a decision approach consistent with the framework of Chapter 6.

Here in Chapter 7 we attempted to apply decision analysis to a siting study in Sussex County, New Jersey. The analysis was a first cut. What was learned is that decision analysis can be useful if the analysts have the required expertise. The attempt to come up with "hard numbers" for the various sites brought home the realization that fairly in-depth analysis of *each* site is required if evaluation of sites is to be meaningful. Most siting studies carry out a rather superficial site analysis—arguing that to do otherwise is time and cost prohibitive. Such superficiality may make their results rather meaningless.

Continued work is required to implement the framework of Chapter 6 into an approach that all parties—analysts, decision makers, and the public—feel comfortable with.

CHAPTER SEVEN REFERENCES

BASS, J. M., Memorandum, "A Comparison of Conjoint Analysis and Decision Analysis," Arthur D. Little, Inc., Cambridge, MA, 9 July 1982.

CONVERSE/TENECH, GEOTECHNICAL AND ENVIRONMENTAL CONSULTANTS, *Identifying Potential Sanitary Landfill Disposal Sites/Systems for Sussex County, New Jersey,* prepared for the Sussex County Board of Chosen Freeholders, November 1981.

HUNT, C. B., *Natural Regions of the United States and Canada,* W. H. Freeman and Company, Philadelphia, 1974.

KEENEY, R., *Siting of Energy Facilities,* Academic Press, New York, 1980, 413 pp.

———, "The Art of Assessing Multiattribute Utility Functions," *Organizational Behavior and Human Performance,* Vol. 19 (1977), pp. 267–310.

——— and E. F. WOOD, "An Illustrative Example of the Use of Multiattribute Utility Theory for Water Resource Planning," *Water Resources Research,* Vol. 13, No. 4 (August 1977).

——— and H. RAIFFA, *Decisions with Multiple Objectives,* John Wiley & Sons, New York, 1976.

RAIFFA, H., *Decision Analysis,* Addison-Wesley Publishing Co., Reading, MA, 1965.

APPENDIX A | HYDROGEOLOGIC PARAMETERS FOR SITING LAND EMPLACEMENT FACILITIES

A.1. INTRODUCTION

This appendix describes the hydrogeological parameters that influence the evaluation and ranking of alternative land emplacement facilities. If we assume that present-day performance of "secure" landfills may be inadequate and leakage may occur, then hydrogeological parameters become important when we site a land emplacement facility for hazardous waste. Optimization of those hydrogeologic criteria that minimize contaminant movement provides a natural system to retard pollutant travel and attenuate any leakage. This additional barrier may be needed to mitigate the movement of a plume of contaminants before it pollutes a present or potential water supply. Which hydrogeologic criteria, then, are important in siting a land emplacement facility for hazardous waste for an extended period?

Derivation of the critical hydrogeologic parameters that minimize contaminant migration from a land emplacement facility is an iterative process that involves many steps. We need to understand the characteristics that influence pollutant travel time and to determine which ones are more important than others.

The siting problem can be divided into four main areas: geology, soil, hydrology, and climate. *Geology* is the interdependence of the structural, stratigraphic, and physiographic characteristics of rocks. Bedrock geology determines the structural framework that surfaces as landforms — the more resistant geological formations form ridges, while those more susceptible to erosion tend to form low regions or sometimes valleys. *Soil* is defined as the mantle of weathering between the atmosphere and unweathered rock. The major characteristic of soil transport is *capacity,* which influences the travel time of a pollutant.

Hydrology deals with the properties, distribution, and circulation of water. Both groundwater and surface-water supplies have the potential to become polluted. The proximity of a site to water supplies and the nature of materials that occur between a site and a water supply influence contaminant migrations. These changes in topography and hydraulic conductivity of a geological formation influence surface and groundwater flow.

Climate is considered a driving force in contaminant migration, but we may exclude it when considering potential sites within the same region, where climate is unlikely to vary significantly.

A.2. GEOLOGY

The first parameter to be considered is the geology of a potential site for the land emplacement facility. Geology will be defined as the interdependence of the structural, stratigraphic, and physiographic characteristics of rocks. It will be divided into the two subparameters of local topography and the quality of host rock, both of which have numerous subdivisions that can be used to measure the performance of a site.

A.2.1. LOCAL TOPOGRAPHY

Natural and physical features define the local topography. The lay of the land will be considered in terms of its nearness to waterbodies, variation in elevation, slope, intervening terrain, and rock outcrops.

Nearness to Surface Water Bodies: A surface water body such as a stream or lake is the intersection between the groundwater and the surface. If some contaminant should leach from the site and travel to a nearby stream, the plume is more likely to flow downstream than to spread beyond it. Distance from site to stream directly influences the contamination travel time — the longer the travel distance, the longer the time needed for a contaminant to reach the natural boundary. Stream density can influence the content of groundwater contamination. For example, when streams are thinly scattered, the distance between them naturally increases. Therefore, contaminants from a site will have a longer travel time in the ground, where they may be sorbed and reacted with the soil, leaving fewer pollutants to reach the stream. On the other hand, closely spaced streams decrease the ground contamination zone at the expense of stream contamination.

Variation in Elevation: The elevation of a particular site is not important by itself, but it becomes significant in relation to the surrounding area. Sites with high elevation relative to surrounding areas often have a high potential for groundwater movement away from the site. Groundwater flows from areas of high elevation (potential) to low. Thus a site's relative elevation often determines the distance that the contaminants will eventually travel before reaching a stable location, where relatively low potential occurs. The higher the initial potential, the greater the distance a contaminant must travel and the greater the polluted area, as shown in Figure A.1.

The underlying geology, such as rock type and structure, heavily influences landforms and in turn the variation in elevation. *High topographic relief* means an area of high hills or deep valleys. When such a region has low permeability, it generally has a steep hydraulic gradient and a shallow water table. When the permeability is good to moderate, it usually has a low hydraulic gradient and a deep water table. Both permit little time for sorption of contaminants by the soil. On the other hand, *low topographic relief* has typically a shallow water table, which also means slow groundwater movement and little time for attenuation of contaminants.

Slope: Slope refers to any ground whose surface forms an angle with the plane of horizon, whether a natural or an artificial incline such as a hillside or terrace. Since water flows from a place of high to low potential, slope influences both surface and groundwater flow movement. Therefore, as ground-surface slope increases, a more rapid decrease in elevation head occurs, causing an increase in velocity and discharge. How does this affect a land emplacement facility site?

A facility at the top of a hill receives little or no runoff except from direct rainfall. In contrast, a facility in an area at the bottom of a hill will collect runoff from the surrounding region in addition to its direct rainfall. The strength requirements are less for a facility built on a flat slope than on a steep one. Therefore, construction of such a site is probably easier and less

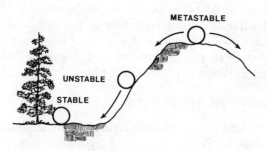

Figure A.1.
Equilibrium schematic.

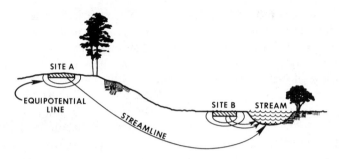

Figure A.2.
Schematic of streamlines and equipotential lines for two sites.

expensive. A facility between the two locations will have to deal with runoff from an area upslope and the runoff from the facility downslope. A high-elevation site may have the advantage of remoteness from the water table, but the disadvantage of contaminating a larger soil zone, as illustrated in Figure A.2. Thus, tradeoffs must be made between sites with high and low slopes.

Intervening Terrain: The intervening terrain is the stretch of ground and surface features between a land emplacement facility site and a designated boundary such as a stream. Surface features include intervening water bodies, vegetation, local topography, and rock outcrops. Vegetation, for instance, slows down surface runoff by substantially decreasing surface runoff from lands down to the site and by increasing the chance for water to infiltrate into the ground. Therefore, if a site is constructed in a low region, runoff can be slowed by planting vegetation upslope, causing less water to flow over and percolate through the contaminated area to the water table.

Rock Outcrops: An outcrop is a piece of stratum that surfaces at the ground and tends to be more resistant than surrounding strata that have already eroded away. Study of its components and structural characteristics, combined with a little knowledge of the surrounding region, can yield a more detailed geological history. However, if a rock outcrop is in a potential site for facility, it will restrict site development and will have to be removed up to an approximate 30-foot depth. If use of heavy equipment is limited, rock outcrop(s), even in an area predominantly soil, may be difficult to remove. This may lead to excessive construction costs.

A.2.2. QUALITY OF HOST ROCK

The quality of host rock may be measured by its structural integrity and regularity of deposit. Bedrock geology determines the structural framework that surfaces as landforms. Hunt (1974) divides the United States structurally into 34 natural regions, called *physiographic provinces,* and groups them into eleven major divisions. For example, New Jersey is comprised of the Continental Shelf and Coastal Plains.

Structural Integrity: Structural integrity of host rock will be discussed in terms of seismic risk zones, dipping and cleavage, inconsistencies, weathering, and mineral resources.

Seismic Risk Zones: A hazardous waste siting report by the Delaware River Basin Commission refers to the Modified Mercalli Intensity Scale and defines seismic risk zones as "areas where an earthquake of Modified Mercalli VII has occurred." Previous earthquake activity is significant because faulting could have structurally weakened the surrounding region through seismic settling, landslides, liquefication, and fracture development. The presence of faults and fractures is extremely important because they provide a natural pathway for flow of contaminants, even in low-permeability and low-porosity rock. Future seismic activity could damage landfill cells and storage tanks of a land emplacement facility during or after the construction, filling, and closing of site, unless it is structurally designed to withstand ground motion.

Regulation under Section 264.18(a) of the Resource Conservation and Recovery Act prohibits location of a new land emplacement facility within 200 feet of any fault. The standard is specified in terms of distance from a fault rather than fault zones, because geologic studies show that deformation is greatest nearest a fault and usually negligible beyond 300 feet. Therefore, 200 feet is considered an adequate distance for a siting boundary.

Dipping and Cleavage: The *dip* of a host rock measures the plunge of its beds. Tilting of beds may influence the hydraulic gradient. For example, a clay layer with low permeability may act as a barrier through which water does not flow readily, thereby confining the lower aquifer. Less force is needed for bedding planes to slip across one another than to fracture across strata; therefore, waterpaths parallel to bedding planes are more likely to occur.

Cleavage is a deformation by shearing, but primarily by flattening along planar bed surfaces, causing them to split thinly apart. When the beds are parallel to axial-plane stress, so too is the cleavage. If parallelism did not exist, however, shear may be oblique, causing more noticeable cleavage in the hinges rather than the flank of a bed. Generally a new mineral, mainly of mica, grows perpendicular to the stress direction. Cleavage indicates the previous existence of a force strong enough to deform the bedding planes and now may provide an easier route for contaminated flow.

Inconsistencies: A host rock may have a significant inconsistency or abnormality, such as an igneous rock extrusion or intrusion, which may influence the flow of water and contaminants. The most common kinds of igneous rocks are granite and basalt. An igneous intrusion, such as a dike, is formed by the solidification of molten rock that was physically injected upward from the earth's center into surrounding rock. An increase in temperature and pressure may alter the structure and mineral composition of surrounding rock, causing them to become somewhat metamorphosed. The igneous intrusion usually fractures during cooling, causing it to become more permeable than the surrounding region, thereby acting as a drain for contaminant flow. In addition, the dike may be composed of different minerals, which could have a different effect on contaminant attenuation, through chemical processes, than the surrounding area.

Another geological inconsistency that must be considered is the presence of aquicludes. An *aquiclude* is a geological formation, consisting of significant amounts of clay, which can alter water movement because of its relative impermeability. An aquiclude may act as a lower or upper boundary of an aquifer. However, often the clay bed will thin out or disappear and provide an interconnection between aquifers. This variable makes prediction of water movement even more complex. The presence of a clay lens may be difficult to predict because of its small size and lack of exposure at the surface.

Weathering: Weathering completes the rock cycle by disintegrating rock sources through physical and chemical processes. Water is a principal means of weathering. In a liquid state, water leaches ions from grain surfaces and enables the growth of plants, whose roots release organic acids that aid mineral decomposition. When water is frozen, it expands and can break even the strongest rocks.

Physical weathering includes frost weathering, diurnal temperature change, sheeting, and biologic factors such as burrowing. *Chemical weathering* is the process by which a system involving rocks, air, and water approaches equilibrium at or near the surface of the earth (Blatt et al. 1980). In most cases, minerals that easily crystallize at depth readily disintegrate when brought up to the surface—in the same order as the Bowen reaction series. The Bowen reaction series states that minerals form in a definite sequence, with iron, magnesium, and calcium silicates (olivenes and calcium feldspars) crystallizing first.

A host rock may become weathered because it was at one time exposed to weathering at the surface before burial and/or the zone of weathering at the surface had penetrated down to the host rock. The extent of weathering in the host rock is important, because it may explain the presence in the rock of fracturing and/or stability of minerals that influence the relative exchanging power of ions with contaminants.

Mineral Resources: Areas of past and present development of mineral resources are significant when we study the quality of host rock. Land subsidence due to subsurface mining could harm the structural integrity of host rock on which a potential site may be constructed. For example, in northeastern Pennsylvania anthracite coal is broadly mined, and the tunneled area would make a weak foundation for a land emplacement facility. Also, the possibility of developing mineral resources in the area of a chosen site should be considered.

Regularity of Deposits: The regularity of a deposit of host rock may be measured by its areal extent, thickness, and depth from surface.

Areal Extent: Speed and direction of movement of contaminants from a pollutant source depend on the medium through which they pass. It is easier to predict migration in a homogeneous isotropic medium, which has a uniform structure and properties that are the same in all directions at a particular point, so that a contaminant will migrate equally in all directions. However, few deposits are totally homogeneous and/or isotropic. An area is desirable whose deposits are regular and large enough for a facility, particularly when structural integrity of the host rock is sound. When a deposit does not cover the area, a facility may have to be designed to deal with the irregularities. This makes construction more complicated and more expensive.

Thickness: The distance traveled by a pollutant through a medium affects the flow period. Therefore, the thicker a deposit, the longer a flow period through the deposit. It is easier to estimate the contaminant flow period for a thick homogeneous deposit than for one composed of numerous thin layers. For instance, alternating sand and clay layers have different permeabilities. Flow through a sand layer is quicker than through a clay layer. This makes prediction more difficult, especially when considering refraction at boundary layers or the possibility of a "wick" effect.

The *wick effect* is a phenomenon in which downward infiltration is slowed at the boundary where unsaturated fine-grained sediments overlie unsaturated, well-sorted coarse-grained sediments. When gravitational forces are greater than capillary forces, vadose water from fine sediments at saturation level drains downward into coarse sediments. A common gardening pot with gravel at the bottom overlain with soil shows this effect. Therefore, by keeping buried wastes away from soil or vadose water through a layer of small pebbles, a natural barrier is created. Winograd (1981) describes a six-year investigation in France at the Centre d'Etudes Nucleaires de Cadarache where no water infiltrated from fine soil into the gravel of an experimental trench in a humid climate, even when 40 cm of water were applied over several hours. The capillary barrier keeps infiltration near the surface to increase rate of evaporation and transpiration.

Depth from Surface: The depth from the surface to host rock is significant in analysis of host-rock quality. Host rock acts as a reference or possibly impermeable boundary surface to which contaminants migrate. The depth may influence host-rock characteristics. If host rock is exposed at or near the surface, it may be susceptible to weathering. For example, a carbonate rock such as limestone is subject to solution activity that weakens its structural integrity.

A.3. SOIL

The second general category to be considered in siting a land emplacement facility is the soil of the region. *Soil* is defined as the "mantle of weathering" between atmosphere and unweathered rock (Blatt et al, 1980). A typical soil is composed of 45 percent mineral matter, 25 percent water, 25 percent air, and 5 percent organic matter (Bridges, 1970). The two major characteristics of soil that we will discuss here are its transport capacity and sorption capacity.

A.3.1. TRANSPORT CAPACITY

Transport capacity refers to a soil's ability to allow movement of water-carrying contaminants. Thus, the greater the soil transport capacity, the greater the migration of contaminants, which is undesirable.

Soil Texture: The texture class of a soil influences its porosity and thus its permeability. Soil texture class is based on the percentages of sand, silt, and clay. The diameters of these relative particles are from 2 to .05 mm for sand, from .05 to .002 mm for silt, and less than .002 mm for clay. A simple field test for texture class is to rub a moistened soil sample between the fingers and estimate relative proportions of sand, silt, and clay. A triangular diagram with 100 percent of each constituent at a corner displays a common classification approach. Table A.1 describes texture classes and field test, and Figure A.3 presents a soil triangle diagram adapted from Bridges (1970).

Texture (fine to coarse) class is significant when siting a facility. As texture of soil materials become finer, the groundwater velocity tends to decrease for the same groundwater potential. For example, a coarse-textured sandy soil provides good drainage when not overlain by fine-grained clay. Such coarse-textured material encourages contaminant migration. In addition, as soil texture becomes finer and clay content increases, grain surface area increases, which increases the soil's sorption capacity.

Porosity and Permeability: A soil consisting of small uniform grains tends to have a dense packing arrangement. This creates a low porosity, because the soil has a low ratio of void volume to total rock volume. *Porosity,* a basic aspect of rock fabric, is the ratio of void volume to total rock volume expressed as a percentage. If the structure of a clay mineral contains water, that water should not be considered as part of the pore space. *Grain size* defines a major difference between mudstones and sandstones. The least

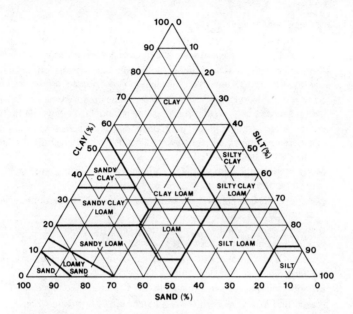

Figure A.3.
Texture class diagram (after Bridges, 1970).

dense packing arrangement of spheres has a porosity of 47.6 percent, while that of the most dense is 26.0 percent.

Hydraulic conductivity measures ease of fluid flow through a porous medium. Darcy's empirical relationship may be manipulated to solve for hydraulic conductivity, if discharge and cross-sectional area (or effective velocity, length of flow, and head loss) are known, as shown in the following equation:

$$K = -V \frac{dl}{dh} \tag{A.1}$$

where

$K =$ hydraulic conductivity with units of length/time,
$V =$ Darcy velocity (effective velocity),
$dl =$ length of flow, and
$dh =$ head loss (energy loss per unit weight).

The hydraulic conductivity for sand is about 10^{-3} cm/sec, while that for marine clay is 10^{-8} cm/sec. In other words, a fluid which flows 1,000 ft/year

Table A.1 Soil Texture Class Descriptions

Sand	Consists mostly of coarse and fine sand and contains so little clay that it is loose when dry and not sticky when wet; when rubbed it leaves no film on fingers.
Loamy sand	Consists mostly of sand, but with sufficient clay to give slight plasticity and cohesion when very moist; leaves a slight film of fine materials on the fingers when rubbed.
Sandy loam	Sand fraction is still quite obvious, which molds readily when sufficiently moist, but in most cases does not stick appreciably to fingers; threads do not form easily.
Loam	Fractions are so blended that it molds readily when sufficiently moist and sticks to the fingers to some extent; it can with difficulty be molded into threads but will not bend into a small ring.
Silty loam	Moderately plastic without being very sticky; a smooth, soapy feel due to the silt is a main feature.
Sandy clay loam	Contains sufficient clay to be distinctly sticky when moist, but the sand fraction is still an obvious feature.
Clay loam	Distinctly sticky when sufficiently moist, and the presence of sand fractions can be detected only with care.
Silty clay loam	Contains small amounts of sand but sufficient silt to confer something of a smooth, soapy feel; it is less sticky than silty clay or clay loam.
Silt	Smooth, silky feel of silt is dominant.
Sandy clay	Plastic and sticky when moistened sufficiently, but sand fraction is still an obvious feature; clay and sand are dominant, and intermediate grades of silt and very fine sand are less apparent.
Clay	Plastic and sticky when moistened sufficiently, and upon rubbing becomes a polished surface; when moist, soil can be rolled into threads and is capable of being molded into any shape, even making clear fingerprints.
Silty clay	Composed almost entirely of very fine material but in which the smooth, soapy feel of silt fraction is modified to some extent by the stickiness of clay.

through gravel will flow .01 ft/year through clay given the same pressure gradient. Hence a contaminant can travel quickly and cover greater distances in a geological formation containing glacial outwash and deltaic sands that are well-sorted sand and gravel beds than in one made up of clay.

Hydraulic conductivity of site may vary within a region because of extensive faulting or fracturing, an aquiclude (see Section A.2.2), or an igneous intrusion or extrusion. It especially varies in the vertical direction because of the sequence of alternating layers of sedimentary soils and the influence of gravity. The direction of the largest hydraulic conductivity is usually parallel to the bedding direction or along fractures. In a region where the formation is homogeneous and covers a large, deep area, prediction of travel time becomes easier and more accurate. For instance, alternating sand and clay layers have different hydraulic conductivities, which makes prediction more difficult — especially in considering the refraction of fluid flow at boundary layers. Construction of a facility in a region where deposits are regular is less complicated than in an area with complex geology. A facility will have to be designed to deal with irregularities such as an aquiclude or an igneous rock that could delude a planner about the quality of a site with local low hydraulic conductivity.

An aquiclude can alter water flow because of its relatively low hydraulic conductivity. While it may act as a lower boundary of an aquifer, one inconsistency or crack may provide an exit for water flow, thus making prediction of water movement difficult. The presence of a thin clay layer is often difficult to predict, owing to an absence of any surface feature.

Permeability and porosity of a site may vary within a region, especially in the vertical direction, because alternating layers of sedimentary rocks (due to depositional sequence) may be only slightly sloped. They are not constant soil properties and may be altered by reaction of contaminants with soil mineralogy or by a change in subsurface stress field or temperature. Mobility of contaminant varies also. In contrast, chromium (Cr), mercury (Hg), and nickel (Ni) are the most mobile and lead (Pb) and copper (Cu) are the least, while others such as arsenic (As), berylium (Be), cadmium (Cd), selenium (Se), and zinc (Zn) depend on conditions. In general, clean sand and gravel without clay or silt tend to be permeable, while clay beds tend to be impermeable. Therefore, a soil with low permeability and porosity can lengthen the flow period and act as a natural defense by retarding movement of contaminants. Glacial outwash plains and deltaic sands are both well-sorted sand and gravel beds with high permeability, thus allowing wastes to move faster and further.

A.3.2. SORPTION CAPACITY

Sorption capacity depends on predominant minerals and pH of soil. Sorption includes both absorption and adsorption of contaminants. *Absorption* is a process whereby the pollutant is assimilated, while *adsorption* is the taking up of gas molecules onto the surface of a particle or its surrounding liquid.

The predominant soil minerals influence interaction between soil particles and contaminants. Their cation exchange capacity (CEC) is extremely important in attenuating heavy metals. The CEC can be defined as the capacity of soil to exchange cations, expressed as the sum for all exchangeable cations. A soil with a high CEC may act as a back-up system for retarding unplanned movement of heavy metals. For example, a soil with an average CEC greater than 30 meq/100 g (milliequivalents per 100 grams of soil) could be considered acceptable, 20 to 30 meq/100 g neutral, and less than 20 meq/100 g unacceptable.

The capacity of soil to retard contaminant migration depends also on the presence of numerous hydrous oxides, particularly iron oxides, and other compounds such as phosphates and carbonates. In particular, heavy metals may be precipitated out of solution, making them unable to travel further.

If iron oxide (FeO) increases and pore water velocity decreases, soil sorption of heavy metals increases.

Crushed limestone limits contaminant movement, especially that of certain heavy metals. It acts as a natural barrier by directly sorbing metal ions, forming less soluble carbonate and calcium compounds and raising infiltrating pH level. The thicker the crushed limestone, the more effective is the liner in retarding heavy-metal migration; experiments with a 2-cm liner show significant retardation (EPA, 1978). For example, at different pH levels the solubility of chromium changes from CrIII to CrVI when the required amount of electron donors is available.

Investigation shows that an equal thickness of combined soil and limestone tends to attentuate more leachate metals than when separated (EPA, 1978). A somewhat permeable soil combined with limestone slows down heavy metals while allowing compounds formed from reactions to pass without clogging the soil's structure, as would occur in soil with a high clay content. A major disadvantage is that limestone's life expectancy is short compared to that of a landfill, making reapplications necessary, because uneven dissolving of limestone liner may create a contaminant exit route.

The hydrogen-ion concentration (pH) of soil influences the dominant removal mechanism for metal cations. It is defined as $pH = \log(1/[H^+])$ and is acid from 0 to 7 and basic from 7 to 14. The dominant removal mechanism for metal cations when $pH < 5$ is exchange or adsorption, and when $5 < pH < 6$, it is precipitation (EPA, 1977). Thus one can consider an average $pH < 6.5$ positive in removal, $pH = 5.0$ to 6.5 neutral, and $pH < 5.0$ negative.

A.4. HYDROLOGY

Hydrology is the science dealing with the properties, distribution, and circulation of water. Proximity to groundwater supplies and to surface water supplies will be considered. Both of these supplies influence contaminant migration and have the potential to become polluted.

A.4.1. PROXIMITY TO GROUNDWATER SUPPLIES

Proximity of a land emplacement facility to groundwater supplies can be evaluated using the distance from the facility (measured from the bottom, for example) to the groundwater table, hydraulic gradient, presence of different aquifers, and closeness to wells.

Distance from the Facility to Groundwater Table: Distance from the facility to the groundwater table may tell the hydrologist something about the region. The groundwater table, or phreatic surface, is the upper surface of a saturated zone where all pore space is filled with water under hydrostatic pressure. As mentioned earlier, a shallow water table (approximately 15 feet or less) tends to be located in a humid region with low permeability. When distance from the surface to the groundwater table is short, contaminant travel time is also short, allowing for little attenuation before pollutants disperse laterally in the saturated zone. Wells may be installed easily and cheaply in a shallow water table to monitor water quality. On the other hand, installation of monitoring wells for a deep water table (usually 75 feet or more) is expensive. It is desirable to have the average distance to the groundwater table large enough so that contaminants may be significantly attenuated and the saturated zone monitored sufficiently. This will permit remedial action to be undertaken if need be. Often such installation can not be accomplished.

Hydraulic Gradient: Hydraulic gradient is the absolute value of the change in hydraulic elevation per unit of length. It influences the direction in which water will flow. A hydraulic gradient sloping away from local groundwater supplies is desired. The steeper the hydraulic gradient, the lower the attenuation time and the faster the water movement. Therefore, a moderate hydraulic gradient may be most acceptable.

Different Aquifers: An aquifer is a geologic formation capable of yielding quantities of groundwater to a well or spring. Depending on local geology, there are often a series of aquifers separated by impervious layers. Their quantity, quality, and use are significant when selecting potential sites.

The quantity of water yielded per unit area per unit drop of piezometric surface is measured by its storage coefficient (S) for confined aquifers and its specific yield (S_s) for unconfined aquifers. Low yield is 10 gallons per minute (gpm) or less, a moderate yield ranges from 10 to 70 gpm, and a high yield is 70 gpm or more. An aquifer's yield may provide an indication of the available groundwater supply.

Quality of groundwater varies depending on geological formations, nearness to the ocean, and previous human activity. For instance, deep groundwater may contain hydrogen sulfide, chlorides, sulfates, and/or carbonates, depending on the type of surrounding rocks. Saltwater intrusion of an aquifer may produce brackish water with a high dissolved salt concentration. Previous human activity may have already degraded a groundwater source, making the supply unusable for human consumption or industrial needs. An aquifer should be considered a precious resource that must be protected from contamination for present and future use, particularly when of good quality.

Use of an aquifer may be determined by its quantity, quality, and local need. For example, a lightly populated area will probably need less water than a power plant. Residents will need pure water to drink and cook with, while a power plant could use, as cooling water, water of a lower quality.

EPA states, in its 1981 Federal Regulations, that a hazardous waste facility is prohibited from locating in the recharge zone of a sole aquifer.

A.4.2. PROXIMITY TO SURFACE WATER SUPPLIES

Proximity to surface water supplies, like proximity to groundwater supplies, involves considerations of quantity, quality, and use. The analogous parameters, such as distance from a land emplacement facility, hydraulic gradient, and stream density, influence contaminant flow from a land emplacement facility to surface water supplies. The difference is that once an aquifer has been contaminated, it is polluted for a long time, whereas a polluted surface supply may be diluted and flushed out quite quickly.

A.5. CLIMATE

Climate is considered a driving force in contaminant migration. For potential sites within the same state, climate is unlikely to vary significantly. For example, the climate of New Jersey is described as rainy with severe winters, a 24-inch average annual snowfall, and a 44-inch average total precipitation. The coldest month averages below freezing and the warmest above 50°F with an average temperature for the year around 50°F. New Jersey's annual mean relative humidity is 70 percent and its free-water evaporation is 34 inches. The latter may also be called the mean annual lake evaporation, which provides an adequate indication of potential evaporation (U.S. Environmental Data Service, 1968).

Site flooding could weaken the structure of a land emplacement facility, causing it to fail and leak wastes. Therefore, it is essential that the facility is not built on a floodplain or area subject to local flooding, unless designed to withstand flood impacts.

Infiltrating precipitation provides a medium in which contaminants may travel. Therefore, stopping possible infiltration by capping the site with clay or by installing a unsaturated barrier will minimize migration of contaminants.

APPENDIX A REFERENCES

BLATT, H., G. MIDDLETON, and R. MURRAY, *Origin of Sedimentary Rocks,* 2d ed. Prentice-Hall, Inc., Englewood Cliffs, NJ, 1980, p. 246.

BRIDGES, E. M., *World Soils,* Cambridge University Press, Cambridge, UK, 1970, p. 8.

ENVIRONMENTAL PROTECTION AGENCY, *Land Disposal of Hazardous Wastes: Proceedings of the Fourth Annual Research Symposium,* EPA A-600/9-78-016, Cincinnati, Ohio, 1978.

————, *Management of Gas and Leachate in Landfills: Proceedings of the Third Annual Municipal Solid Waste Research Symposium Held at St. Louis, Mo., on March 14, 15, 16, 1977,* Cincinnati, Ohio, 1977, p. 14.

HUNT, C. B., *Physiography of the United States,* W. H. Freeman and Company, Philadelphia, 1967, p. 53.

U.S. ENVIRONMENTAL DATA SERVICE, *Climate Atlas of the United States,* U.S. Department of Commerce, Washington, DC, 1968.

WINOGRAD, I. J., "Radioactive Waste Disposal in Thick Unsaturated Zones," *Science,* Vol. 212, No. 4502 (June 26, 1981), p. 1462.

INDEX

Absorption, 155
Adsorption, 2, 22, 155
Anisotropic media, 43
Aquiclude, 32, 150, 155
Aquifer:
　confined, 32
　leaky confined, 33
　phreatic, 33
　unconfined, 33
Aquifuge, 32
Aquitard, 32
Artesian condition, 34
Attenuation, 147
Attribute, 126, 131
　definition of, 117, 122
　　natural, 117
　　surrogate, 117

Bioconcentration factor, 24
Biological degradation, 21
Biomagnification, 23

Cation exchange, capacity of, 155
Chemical carcinogens:
　latency period, 12
　threshold level, 12

Chemical degradation, 21
Chemical reaction, 56
Chemical species, conservation of, 45
Climate, 146, 158
Cohansey Sand, 75–76
Concentration, definition of, 27
Conductivity:
　electromagnetic, 71
　hydraulic, 32, 153–154
Conjoint analysis, 141–143
Conservation, principles of, 35
Conservation-of-mass equation, 39
Continuity equation, 39
Convection, 35, 56
Converse/Tenech:
　consultants, 130
　siting study of, 133–141
Coupled system, 19

Darcy velocities, 43
Darcy's law, 43, 44
Decision analysis, 121, 143
　comparison to conjoint analysis, 143
Deposits, regularity of, 151
Desorption, 22
Difference notation operator, 37

Dilute solutions, 27
Dispersion, 56
Dispersivity, 48
Disposal facilities, 106
Disposal ponds, 89
Dose commitment, 10
Dynamic state, 15

Environmental Protection Agency (EPA), 106
Equilibrium, 15
Equipotential lines, 148
Exponential functions, 17, 26
Exposure concentration, 10

Federal Water Quality Act, 53
Freundlich isotherm, 22

Geology, 145
  bedrock, 59
  in siting (*see* hydrogeology)
  structural integrity, 149
  surficial, 59
Groundwater, 60
  Darcy's law, 42, 43
Groundwater table, 156

Half-life, 18
Hazard assessment, 10
Hazardous waste:
  cost of disposal, 7
  disposal method, 7
  facilities siting, 106–120, 121–143, 145–158
  land disposal of, 2
Host rock, quality of, 148
Hydraulic gradient, 157
Hydraulic head, 33
Hydrogeology, criteria in siting, 137, 146–158
Hydrology, groundwater, 31
Hydrolysis, 21

Industrial waste, disposal ponds, 86
Isotropic media, 43

Kirkwood Formation, 76

Land emplacement facilities (see hazardous waste, facilities siting)
Leachate, 3
Love Canal, containment project, 4

Mechanical dispersion, 36
Model:
  mathematical, 14, 81, 89
  numerical, 30
Modeling:
  definition of, 118
  numerical, 51
  review of, 118
Molecular diffusion, 35
Monitoring network:
  design, 54, 58–69
    Level 1, 55
    Level 2, 55, 57, 61–63
    Level 3, 55, 63–68
  error analysis of, 59
  expansion of, 68
  operation of, 71–73
Monitoring techniques:
  errors in, 73
  geophysical, 70

Numerical models, errors in, 52

Objectives:
  in monitoring network design, 56–59
  in siting facilities, 107, 122
  for Sussex County siting study, 125
Octanol-water partition coefficient, 24

Partition coefficient, 22
PCB's, 8
Permeability, 32
pH, 156
Photodegradation, 21
Photolysis, 21
Physiographic provinces, 148
Piezometric head, 34
Piezometric surface, 34
Pollution transport, modeling of, 35
Price Landfill, 75–76
Pumping, strategies of, 82

Radar, ground-penetrating, 71
Reaction, 35
　first-order, 17
　zero-order, 17
Refractory chemicals, 21
Remedial schemes, 82
Resistivity, electrical, 71
Resource Conservation and Recovery Act
　　(RCRA), 5, 53, 56, 149
　definition of hazardous waste, 6
Retention dose, 11
Risk assessment, 9

Safe Drinking Water Act, 53
Seismic, 71, 149
Sensitivity analysis, 93, 119
Site, impacts on, 126
Site-selection:
　attributes for, 115
　decision analysis in (*see* decision
　　analysis)
　hydrogeological characteristics (*see*
　　hydrogeology)
　objectives for, 114–115
　process, 108–120
　　characteristics, 108–109
　　concerns of, 107
　Sussex County study, 121–143
Slope, 147
Soil, 146, 152
　permeability of, 152
　porosity of, 152
　texture of, 152
Sorption, 21, 147
　capacity of, 155
South Brunswick, 61, 64
Specific yield, 157
Steady-state conditions, 15
Storage coefficient, 42, 79, 157
Stratigraphic boundary, 70
Streamlines, 148
Superfund, 3, 5
Surface Mining Control and Reclamation
　　Act, 53
Sussex County, New Jersey, 122, 127, 129,
　　133
Sussex County siting study, 125–126, 140

Time-variable, 15
Topography, 146
Transformation, 15, 20
Transmissivity, 32, 79
Transport, 15
Transport capacity, 152

Utility function, 122, 126-133
　evaluation of, 127–129
　　values for Sussex County siting study,
　　　141

Variability, 64
Velocities, superficial, 43
Volatilization, 20

Water table, 33
Weathering, 150
Wick effect, 151

Yield coefficient, 18